A Feminist Urban Theory for Our Time

Antipode Book Series

Series Editors: Vinay Gidwani, University of Minnesota, USA and Sharad Chari, University of California, Berkeley, USA

Like its parent journal, the Antipode Book Series reflects distinctive new developments in radical geography. It publishes books in a variety of formats – from reference books to works of broad explication to titles that develop and extend the scholarly research base – but the commitment is always the same: to contribute to the praxis of a new and more just society.

Published

A Feminist Urban Theory for our Time

Rethinking Social Reproduction and the Urban

Edited by

Linda Peake, Elsa Koleth,
Gökbörü Sarp Tanyildiz,
Rajyashree N. Reddy &
darren patrick/dp

WILEY

Registered Office(s)
John Wiley & Sons, Inc., 111 River Street, Hoboken, NJ 07030, USA
John Wiley & Sons Ltd, The Atrium, Southern Gate, Chichester, West Sussex, PO19 8SQ, UK

Editorial Office
9600 Garsington Road, Oxford, OX4 2DQ, UK

For details of our global editorial offices, customer services, and more information about Wiley products visit us at www.wiley.com.

Wiley also publishes its books in a variety of electronic formats and by print-on-demand. Some content that appears in standard print versions of this book may not be available in other formats.

Library of Congress Cataloging-in-Publication Data
Names: Peake, Linda, 1956- author.
Title: A feminist theory for our time : rethinking social reproduction and the urban /
    Linda Peake, Elsa Koleth, Gökbörü Sarp Tanyildiz, Rajyashree N. Reddy & Darren Patrick.
Description: Hoboken, NJ : John Wiley & Sons, 2021. | Series: Antipode book series |
    Includes bibliographical references and index.
Identifiers: LCCN 2020058509 (print) | LCCN 2020058510 (ebook) | ISBN 9781119789147
    (hardback) | ISBN 9781119789154 (paperback) | ISBN 9781119789185 (pdf) |
    ISBN 9781119789178 (epub) | ISBN 9781119789161 (ebook)
Subjects: LCSH: Feminist theory. | Queer theory. | Sociology, Urban.
Classification: LCC HQ1190 .P43 2021 (print) | LCC HQ1190 (ebook) | DDC 305.42--dc23
LC record available at https://lccn.loc.gov/2020058509
LC ebook record available at https://lccn.loc.gov/2020058510

Cover image: © Africa Studio/Shutterstock
Cover design by Wiley

Set in 10.5/12.5 and SabonLTStd by Integra Software Services, Pondicherry, India
Printed and bound by CPI Group (UK) Ltd, Croydon, CR0 4YY

C115401_190721

Contents

List of Contributors

James Angel
Department of Geography
King's College London
London, UK

Natasha Aruri
K LAB
Institute for City and Regional
Planning
Berlin

Belinda Dodson
Department of Geography
Western University
London, ON
Canada

Emily Fedoruk
College of Liberal Arts
University of Minnesota
Minneapolis
Minnesota, USA

Friederike Fleischer
Department of Anthropology
Universidad de los Andes
(Columbia)

Bogota
Columbia

Tom Gillespie
Global Development Institute
The University of Manchester
Manchester, UK

Kate Hardy
Division of Work and Employment
Relations
University of Leeds
Leeds, UK

Meera Karunananthan
Blue Planet Project
Ottawa, ON
Canada

Mantha Katsikana
Department of Geography
York University
Toronto, ON
Canada

Elsa Koleth
The City Institute
York University
Toronto, ON
Canada

Faranak Miraftab
Department of Urban & Regional
Planning
University of Illinois, Urbana-
Champaign
Champaign, IL
USA

Camila Esguerra Muelle
Interdisciplinary Group on Gender
Studies
National University of Colombia
Bogota
Columbia

Beverley Mullings
Department of Geography and
Planning
Queen's University
Kingston, ON
Canada

Diana Ojeda
Centro Interdisciplinario de
Estudios Sobre Desarrollo
Universidad de los Andes
(Columbia)
Bogota
Columbia

darren patrick/dp
Women and Gender Studies
Institute
The University of Toronto
Toronto, ON
Canada

Linda Peake
Faculty of Environmental
and Urban Change
York University
Toronto, ON
Canada

Rajyashree N. Reddy
University of Toronto
Scarborough
Toronto, ON
Canada

Liam Riley
Balsillie School of
International Affairs
Wilfrid Laurier University
Waterloo, ON
Canada

Susan Ruddick
Department of Geography
The University of Toronto
Toronto, ON
Canada

Nathalia Santos Ocasio
Department of Geography
and Planning
Queen's University
Kingston, ON
Canada

Gökbörü Sarp Tanyildiz
Department of Sociology
Brock University
St. Catharines
Brock, ON
Canada

Series Editors' Preface

The *Antipode Book Series* explores radical geography 'antipodally', in opposition, from various margins, limits or borderlands.

Antipode books provide insight 'from elsewhere', across boundaries rarely transgressed, with internationalist ambition and located insight; they diagnose grounded critique emerging from particular contradictory social relations in order to sharpen the stakes and broaden public awareness. An *Antipode* book might revise scholarly debates by pushing at disciplinary boundaries, or by showing what happens to a problem as it moves or changes. It might investigate entanglements of power and struggle in particular sites, but with lessons that travel with surprising echoes elsewhere.

Antipode books will be theoretically bold and empirically rich, written in lively, accessible prose that does not sacrifice clarity at the altar of sophistication. We seek books from within and beyond the discipline of geography that deploy geographical critique in order to understand and transform our fractured world.

Vinay Gidwani
University of Minnesota, USA

Sharad Chari
University of California, Berkeley, USA

***Antipode* Book Series Editors**

Preface

As feminist, Marxist, postcolonial and queer scholars, our concern in this book is to show how social reproduction is foundational in comprehending urban transformation. Social reproduction is, of course, not just an analytical framing but also an organizing call for feminist scholars and our contention is that if we want an urban theory for our time, it needs to be feminist. Feminism is not simply a 'discipline', 'theory', or 'ideology', but a worldview, a lived praxis that provides a platform for engaged analysis.

The book's origins lie in our belief in the necessity of feminist urban knowledge production, a belief further endorsed by our prior critical engagement with the analytical framework of planetary urbanization and our collective ruminations during and post this engagement on the nature of urban theory (Reddy 2018; Ruddick et al. 2018). Not least the considerable response to the theme issue of *Society and Space* (Peake et al. 2018) showed us that there was an audience desirous of troubling the hegemony of urban theory. Moreover, our approach of working as a team across hierarchies of junior and senior scholars, generations, genders, sexualities, institutions, and disciplines – a praxis we refer to as 'the intergenerational social reproductive labor of knowledge production' (Peake et al. 2018, p. 377) – had been fruitful and positive and we wanted it to continue. It was as much a pedagogical experience of reading and writing together, and sharing meals, as it was an exploration of our places within the academy and an intellectual foray into urban theory. And while Roza Tchoukaleyska left for Newfoundland, Elsa Koleth, a new post-doctoral fellow at the City Institute at York University, joined us.

Our Canadian location, although mediated by our own migrations from Australia, India, Turkey, the United Kingdom, and the United States, led us to put out a call for papers in 2018 on the theme of a 'feminist urban theory for our time' at the annual conferences of the Canadian

Association of Geographers, the American Association of Geographers, and the Urban Affairs Association. Some of our contributors answered these calls for conference papers, while others are members of the Gen-Urb (SSHRC-funded) project (Urbanization, gender and the global south: a transformative knowledge network). Much has been written in urban scholarship by feminist and postcolonial scholars on global circuits of knowledge production and the privileging of Anglo-American scholarship. We recognize that sending conference calls to those attending North American based conferences not only reduces the geographic locations of the research reported but also heightens whiteness as a political and epistemological position and that this volume is thereby limited in its capacity to pluralize and broaden the epistemic community engaged in feminist urban theory. Nonetheless, our authors come from Colombia, Germany, Greece, Iran, Palestine, Puerto Rico, and South Africa as well as Canada, the United States, and the United Kingdom. And they report on research based in cities in Argentina, Canada, Columbia, Greece, Haiti, Indonesia, Kenya, Palestine, Puerto Rico, Spain, the United Kingdom, the United States, Zambia, and Zimbabwe.

Collectively, the contributors explore how the urban can be understood through the light shone on the dynamics of social reproduction in people's everyday lives and their interaction with processes of capitalist accumulation as they are actively reconfigured through the manifold processes of contemporary urbanization. They proffer the insight that a feminist social reproduction approach to the urban offers not only an engaged analysis of the variegated nature of the urban but also of the relationship between capitalism and the production of social difference. With a focus on the everyday urban contexts within which social reproduction takes place, the various contributions make visible the insidious, often unacknowledged, and seemingly innocuous ways in which lives are being transformed, highlighting the moral economies within which these contexts are normalized and rendered ordinary rather than unlivable.

As we wrote, the pandemic conditions that have gripped the globe in the most catastrophic and intimate ways have cast many of the processes discussed in this book into sharp and brutal relief. We are reminded once again of the absolute necessity of social reproduction for human survival, of the fragility of the infrastructures and bodies that make social reproduction possible, and of the grossly unjust systems of power that secure the social reproduction of the few through the disposability, expulsion, and annihilation of many others. Whether seen through the near collapse of health and welfare systems in urban centres already ravaged by austerity, the mass exodus of impoverished migrant workers back to rural villages, or the significant reductions in women's participation in numerous labour forces, the prevailing crises of social reproduction around the world have been exacerbated exponentially in current conditions.

Finally, we would like to acknowledge the work undertaken by the contributors to this book, a number of them junior scholars, as well as their patience in revising various drafts of their chapters. We also thank *Antipode*'s book series editors, Sharad Chari and Vinay Gidwani, who started the process of creating this book with us and Nik Theodore who saw us through to the end, for their interest in our project and providing us with the opportunity to pursue it through to its publication.

Linda Peake, Rajyashree N. Reddy, Gökbörü Sarp
Tanyildiz, Elsa Koleth, and darren patrick/dp

References

Peake, L., Patrick, D., Reddy, R., Tanyildiz, G., Ruddick, S., and Tchoukaleyska, R. (2013). Placing planetary urbanization in other fields of vision. *Environment and Planning D: Society and Space* 36 (3): 374–386. doi: 10.1177/0263775818775198

Reddy, R.N. (2018). The urban under erasure: Towards a postcolonial critique of planetary urbanization. *Environment and Planning D: Society and Space* 36 (3): 529–539.

Ruddick, S., Peake, L., Patrick, D., and Tanyildiz, G.S. (2018) Planetary urbanization: An urban theory for our time? *Environment and Planning D: Society and Space* 36 (3): 387–404. doi: 10.1177/0263775817721489

1

Rethinking Social Reproduction and the Urban

Gökbörü Sarp Tanyildiz (Brock University)
Linda Peake and Elsa Koleth (York University)
Rajyashree N. Reddy (University of Toronto Scarborough)
darren patrick/dp and Susan Ruddick
(University of Toronto)

Introduction

As we move through the 21st century, the changing geographies of urbanization, increasingly unfettered capital accumulation, unprecedented levels of migration, and crises of climate and viral pandemics, have added further urgency to the seemingly intractable question of which categories and methods are adequate to understanding and researching the urban. And yet, notwithstanding their increasing inability to explain 21st century urbanization and urbanism in their 'infinite variety', the 19th and 20th century economic compacts upon which mainstream and Marxist urban theory have been based – the nexus of urban land, circuits of capital, production, and agglomeration economies – remain in place. While it is still customary to approach the contemporary urban by recounting the shifts in the structures and agendas of capitalism and the impacts of these shifts on daily life, we contend it is not possible to think through the urban without considering the role and relations of social reproduction: which are neither

A Feminist Urban Theory for our Time: Rethinking Social Reproduction and the Urban,
First Edition. Edited by Linda Peake, Elsa Koleth, Gökbörü Sarp Tanyildiz, Rajyashree N.
Reddy & darren patrick/dp.
© 2021 John Wiley & Sons Ltd. Published 2021 by John Wiley & Sons Ltd.

subordinate to production, nor an embellishment; neither something to be 'added to urban theory', nor an after-effect to the analysis of processes of urbanization that was assumed adequate without it. Notwithstanding the ubiquity of the global crisis in social reproduction, large swathes of mainstream and critical urban scholarship continuously fail to recognize both the analytical interdependence between relations of social reproduction and production, and how this interdependence shapes social relations and urban futures. It has been left to feminist urban scholars, time and again, to call attention to the radical incompleteness of urban thought, decrying theory that writes life and lives out of time and place[1] (see, for example, Kollontai 1977 [1909]; Burnett 1973; Hayford 1974; Lofland 1975; Mackenzie 1980; Markusen 1980; Wekerle 1980; Hayden 1983; Ferguson *et al.* 2016; Fernandez 2018; Kanes Weissman 2000; Rendell, Penner, and Borden 2000; Spain 2002; Mitchell, Marston, and Katz 2004; Meehan and Strauss 2015; Miraftab 2016; Peake 2017; Pratt 2018; Ruddick *et al.* 2018). We offer this book with the hope that is amplifies and resonates with this long-growing feminist chorus.[2]

Our central problematic, then, is to ask how social reproduction might generate different ways of knowing and investigating the urban in its constitutive and regulative relations. We have argued previously (Peake *et al.* 2018; Ruddick *et al.* 2018) that in terms of their social ontology, urban geographies are geographies of living, yet urban theories have distilled this living to capital and wage-labour in processes of production. This filtering process is part of the hierarchical knowledge production in which the knower's positionality is integral to theorizing and valuing subjectivity and experience. The dominant Enlightenment-bequeathed academic knowledge system of phallogocentrism – the privileging of determinateness and of the masculine (Derrida 1978; Cixous and Clément 1986 [1975]) – has sundered economic production and social reproduction, pitting them against each other as dichotomous opposites and privileging economic production over social reproduction. The dominant urban epistemology is thus one in which economic production and social reproduction have been historically presented as separate and different, both geographically and analytically, signifying domination and subordination, greater and lesser value, respectively. We start however from a problematic of the constitution of economic production and social reproduction as inseparable; they are two dimensions of one integrated system that are constructed, temporally and spatially, in processual relation to each other and marked by differentiation and struggle. We also start not with a notion of being 'different', but with social difference, which we understand conceptually as 'relational connectedness' (see Ware 1992, p. 119), whereby colonial, patriarchal, racist, and heteronormative disciplinary systems of domination and oppression play out through processes of production and social reproduction, attempting to determine who has the right to belong and the right to life itself. Finally, we argue that the potential for urban transformation lies both in the small slippages and seemingly prosaic aspects of everyday life, as well as in more exceptional events and encounters, organized and spontaneous struggles, and in the supplemental space of undecidability and indeterminacy.

The process of the urban coming into being through the relational connectedness of social reproduction and production is thus never fully complete. Only partially determined, this urban process is exceeded both by the struggle of contending classes within capitalist history, including its present, and by the social and political relations that reverberate within histories that can neither be sedimented as, nor absorbed by, the history of capitalism and its attendant structures of subjectivity. We argue that the enduring necessity of social reproduction constitutes an embodied openness to these different histories, an openness that is violently truncated by hegemonic regimes of exploitation and oppression. Tapping into this openness through the urban everyday, we can unsettle the apparent certitude of capitalist value-producing logic and its historical teleology. The urban, therefore, not only spatially conditions and mediates the unfolding of the capital-labour contradiction but it is also reshaped and reorganized in this process. Perhaps most importantly for our time, the spatial organization of embodied urbanization is open both to resurgent histories that resist the economy's subsumption of life and to everyday struggles that make other lives and futures possible. These too often ignored aspects of the urban come into focus in this book – an urban that opens to radical histories and struggles of life-making through social reproduction, and a social reproduction that is not an end in itself, but a methodological entry point into understanding how people in their everyday lives shape and reshape the spatial forms of their lives.

How then do we understand social reproduction? First, we consider social reproduction as a real object of knowledge – that is as a conceptually generative construct and productive way of knowing the urban, and of understanding how urbanization is being reorganized and resisted. Writing amidst the vestiges of modernity – of people-making, public space, freedom, citizenship (already profoundly limited forms) – that are all but eroded, we ask how people's agency, struggles, desires, hopes, and dreams, might be rethought in light of the erasure of the social wage and social contracts and their replacement by demonization, dispossession, and the downloading of responsibility for social life to 'the individual'. This increasing precarity of urban life and how this life is reproduced, in conjunction with the analytical framings used to examine them, has put the feminist intellectual and political project of social reproduction back on the urban agenda with a new urgency, engendering praxis and producing ideas that can be socially and epistemologically transformative (see, for example, Teeple Hopkins 2015; Buckley and Strauss 2016; Andrucki *et al.* 2018; Chattopadhyay 2018; Peake *et al.* 2018; Winders and Smith 2019).

Second, notwithstanding the decisive role of social reproduction, it has formed only the theoretical 'constitutive outside' of the urban since non-feminist urban theorizing began (as the 'illegible domain that haunts the former domain as the spectre of its own impossibility, the very limit to intelligibility' Butler 1993, p. xi) (Peake 2016; Roy 2016b; Jazeel 2018; Ruddick *et al.* 2018). We ask how we can transition from treating social reproduction as a mere constitutive outside to being constitutive of how, where, when, and through whose labour the urban emerges. Hence, we see social reproduction as a real object of the urban – an empirical reality to be mapped, documented; a tableau that writes the urban even as it is written by it. Moreover, we consider the who, where, when, and how of social reproduction and the alternative social and spatial relations it produces to be historically contingent and only partially discernable through their specific relationship to the mode of production in which they are unfolding.

Third, our problematic also speaks directly to the imperative to decolonize feminist urban knowledge production, which is not free of hierarchical and imperial thought, produced within a social ontology shot through with whiteness and specific Western ideologies, values, and experiences. It is with this concern of decolonizing the epistemologies and ontologies of existing social reproduction analytical frameworks that we propose *social reproduction qua method* (Tanyildiz 2021), as a tool to think through the relationship between ontology and epistemology, which orients us towards how social reproduction is undertaken. As method, social reproduction is an attempt to explicitly connect some of the main aspects of critical feminist epistemologies – such as emphasizing the locatedness and partial nature of knowledge production and a willingness to continually scrutinize categories of analysis, embedded as they are in specific spatialities and temporalities – to feminist considerations of social ontology (cf. Ruddick *et al.* 2018). Foregrounding what social reproduction can do as an organizing lens at least partially frees us from predetermined sets of implicitly white and explicitly economically reductive analytical categories, providing a much-needed epistemological reflexivity. Such an intentionally open framework enables us to attend to the range of ways in which people shape the circumstances of daily life in relation to conditions of hegemonic capitalist production. This framework not only reveals how capitalist value-producing labour is predicated upon social reproductive labour – thereby providing a more robust analysis of the capitalist mode of production in its totality – it also moves us closer to understanding how the teleological philosophy of history put forward by the proponents of capitalism (and reproduced by capitalist social relations) is only rendered possible through the everyday constriction of a host of other histories and the social relations and subjectivities that can organize life differently. Social reproduction as method is useful then because it does not require us to invest in a specific epistemology and ontology, thereby recognizing the necessity for other epistemologies and ontologies in the conversation.

Expounding social reproduction as method requires elaboration of the relationship between social relations and the relations of social reproduction, as both separate and in relation to each other. Social ontology does not ask 'what is' as classical ontology does. What social ontology does is to investigate the conditions of the possibility of society, the social, and social relations. Put differently, it orients us towards examining the reality of society, the social, and social relations in a formative and integrative fashion. Social reproduction, on the other hand, provides us with the omitted underbelly of society, the social, and social relations. For instance, it shows us: how capitalism (despite its seeming omnipotence) cannot reproduce itself in a capitalist fashion; how capitalism (despite its constantly discarding people out of the wage-labour relation into the reserve army of labour) needs those very 'disposible' peoples for its futurity; how this reveals that (despite patriarchy, white supremacy, and other forms of oppression) women, people of colour, and other oppressed subjects are absolutely essential for the survival of society; and, therefore, how resistance and struggle for the liberation of these peoples are necessary for a better world. What social reproduction does is to give a fuller, more wholesome picture of the society we live in (Tanyildiz 2021). Such rethinking moves us away from considering social reproduction as a unitary theory of oppression towards comprehending it as a method that accounts for the historicities and spatialities of its variegated mobilizations, organizations, and praxes of the particular investigation under consideration. At the same time, forwarding social reproduction as method ensures that social reproduction does not assume another untethered epistemological salience and autonomy.

Social Reproduction

Most conceptualizations of social reproduction and its relationship to capitalist production, especially those within the field of feminist political economy, are derived from Marx's use of the notion (1993 [1885]). Cindi Katz's (2001, p. 711) now iconic understanding of social reproduction as the 'fleshy, messy, indeterminate stuff of everyday life' is deliberately broad and imprecise, as is its conception as 'life's work' (Mitchell, Marston, and Katz 2004). Other definitions, still laid out in broad brush strokes, are more cut-and-dry, along the lines of social reproduction as 'the process by which a society reproduces itself across and within generations.'[3] Yet others have had a preference for more detail. For instance, Brenner and Laslett's (1989, pp. 382–383) now 30-year old definition of social reproduction is still much repeated:

> the activities and attitudes, behaviors and emotions, responsibilities and relationships directly involved in the maintenance of life on a daily basis, and intergenerationally. Among other things, social reproduction includes how food, clothing, and shelter are made available for immediate consumption, the ways in which the care and socialization of children are provided, the care of the infirm and elderly, and the social organization of sexuality. Social reproduction can thus be seen to include various kinds of work – mental, manual, and emotional – aimed at providing the historically and socially, as well as biologically, defined care necessary to maintain existing life and to reproduce the next generation.

It is feminist critiques of classical Marxism as well as feminist political economy analyses of social reproduction's defining relations and categories – labour, work, home, gender, race, class, sexuality, the family, life, and value – that have led to the de-naturalization and problematization of social reproduction. In 1969, a century after the publication of Marx's *Capital*, Margaret Benston (1969) published an article entitled 'The political economy of women's liberation' in the *Monthly Review*. For Western feminism, Benston's pioneering piece placed 'the politics of women's liberation within an anti-capitalist framework' and identified 'domestic labor as the material basis of women's structural relation to capitalist production and their subordination in society' (Federici 2019). In doing so, Benston helped to inaugurate the field of the political economy of gender. The following decade saw a proliferation of work in this area of socialist feminism, which re-envisioned critical political economy as feminist political economy by opening its categories to epistemological scrutiny.[4]

Socialist feminist political economy's most important contribution was the concept of social reproduction.[5] A number of feminist scholars made important and wide-ranging contributions demonstrating that capitalism cannot reproduce itself capitalistically; rather, it downloads the burden of its own reproduction onto women in the form of unwaged work. This was an invaluable insight into how capitalism as a system of private property and exploitation worked in tandem with patriarchy, even though there was no agreement as to the actual nature of this relationship between these two systems of exploitation and oppression. The centrality of the concept of social reproduction, however, was so accepted and uncontested that it became synonymous with the field itself, coming to be known as social reproduction feminism (Ferguson 2020). Not only did this field gender classical Marxist political economy's focus on production, but it also expanded conceptualizations of the modes of production, as well as historicizing and spatializing patriarchy, paving the road towards a more unitary theory of oppression.

In these earlier studies of the role of women's domestic labour in the renewal of labour-power and non-workers, such as children, youth, and adults out of the workforce, the household as the socio-spatial unit of social reproduction was privileged. Contemporary feminists have moved beyond household-based analyses, investigating other sites and modalities of social reproduction, such as those of day care centres, schools, institutions of higher education and training, recreation centres, health centres, and hospitals. These studies were combined with those that explored the ways in which the relations of production are recreated through the inter-generational transmission of material, emotional, and affective resources, including through the nurturing of individual characteristics such as self-confidence, and the establishing of group status and inequality, such as through access to education. Intermeshing with these studies were those that encompassed human biological reproduction centering particularly on childbirth and the obligation of maintaining kin networks and relationships, such as those ordained by marriage, and thus the study of the social organization of fertility and sexuality (Kofman 2017) as well as social constructions of motherhood (Bakker 2007). More recently, scholars in the field have recognized that bonds of care are a central ethic and need within social reproduction, including nurturing in ways that keep people psychically, emotionally, and mentally 'whole'. Social reproduction is, thus, heavily implicated in subjectivity formation in that it comprises the embodied material social practices of those engaging in both the material and emotional activities and relations that bring everyday life into being.

While the activities and relations of social reproduction in these studies have been prescribed and overdetermined as women's work, this has been an exercise fraught with omission, not least in circumscribing who counts as 'woman'. We concur that in many parts of the world women, whether in conjunction with the state, private sector, other family or community members, or on their own, are still central to processes of social reproduction that maintain human life – those that either must be done if people are to survive, or those that lead to improved living conditions or a greater sense of well-being. The epistemological turn of moving beyond the household has enabled a reorientation of social reproduction to the global capitalist system at large and to the multifarious ways in which the renewal of labour-power occurs, such as (ironically) through an increased engagement in the social reproduction of other households via intra- and trans-national migration by nurses, teachers, and live-in caregivers and the flows of remittances these migrants send back to their families. This expanded gaze has led to an increasing recognition that not all women participate in social reproductive work, at the expense of embodied others who do, most commonly across classed and racialized lines, and that other marginalized groups – for example, children, refugees, immigrants, modern-day slaves – regardless of gender, are also heavily engaged in such work.[6]

It is also the case that while embodiment has been a presupposition for the labour engaged in processes of social reproduction (and production) it is increasingly no longer a prerequisite. The costs of the social wage constitute a drain on the production of surplus value (especially shareholder profits). Capital's retreat from the social wage has resulted in the increasing financialization and marketization of social reproduction, assigning it a market value (Bryan and Rafferty 2014). This embodied labour moreover can now be acquired flexibly for select slivers of time, on zero-hour contracts at minimum wages and below. Moreover, artificial intelligence (via various platforms that simulate social interaction) and automation are increasingly supplanting embodied labour. Being stripped of waged employment, the body can be 'employed' as an encasement of desirable parts and organs – such as hair, blood, kidneys – whereby 'biotechnologically isolated, manipulated, and disseminated life is absorbed by capitalist processes' (Floyd 2016, p. 61). For example, biotechnological developments in biological reproduction has led women from being a source of labour–power to becoming a source of living raw material through surrogacy. We understand this multifaceted process of eliminating labouring bodies broadly as a continuation of processes of enclosure.

As the conditions in which social reproduction takes place have become more precarious and attacks upon it have accelerated, its analysis (having fallen into a lull during the 1990s and 2000s) has once again risen to the top of many feminist agendas. With no room for race or other relations of oppression beyond those of class and gender in the early social reproduction analyses, there had been a theoretical divestment, until the most recent revival of social reproduction theory, which brings a rigor to hitherto unaddressed questions (Ferguson *et al.* 2016; Bhattacharya 2017). In the last decade, social reproduction theory has emerged as an attempt to offer a unitary theory of women's oppression. Social reproduction feminists have critiqued earlier feminist political economy analyses for not focusing on 'the multi-faceted complexity of real-world relations and political struggles, as well as the ways in which racial oppression intersects with gendered forms of domination and class exploitation' (Ferguson *et al.* 2016, p. 28). In order to avoid such theoretical fallacies, contemporary social reproduction feminists have reconceptualized their ontological presuppositions in regard to the nature of the social. They argue that relations of oppression that are racialized, gendered, classed, and sexualized, 'are *not* additional systems that just happen to coincide. Rather, they are concrete relations comprising a wider sociality, integral to the very existence and operation of capitalism and class' (Ferguson *et al.* 2016, p. 32). We further add that, to examine the constitutive role of racial difference as a historically sedimented formation, the conceptualization of social reproduction could usefully be brought into conversation with postcolonial urban theory. This is central not only to ensuring that our conceptualization of social reproduction is historicized but, as Ananya Roy (2016a) would argue, is also attentive to historical difference as constitutive of the urban.

Notwithstanding its intimate political and theoretical relations with earlier debates, and sometimes because of this, social reproduction theory is often mistaken as a mere synonym of either domestic labour debates or socialist feminism. And yet it is premised upon distinctive ontological and epistemological propositions in that it foregrounds the internal relationship between capitalist value-producing labour and its often omitted predicate, that is non-capitalistically produced social reproductive labour, by focusing on the latter's necessary but contradictory relation to the capitalist pursuit of surplus value. Through shifting the analytical focus onto this internal relationship, social reproduction theory is able to: historicize the notion of patriarchy vis-à-vis specific modes of production and their attendant social formations; demonstrate that women's oppression is not a pre-capitalist residue that capitalism merely picks up, but is integral to the very logic of capitalism as a system, and is necessarily reinvented as regimes of capital accumulation change; and argue that historically specific forms of patriarchy and capitalism are not external to one another, but, rather, are co-constitutive of each other.

Our understanding of social reproduction builds upon those of social reproduction theorists in that we do not consider it as a coherent stable construct over time and place, but as an historicized and spatialized construct, speaking to multiple layerings, subject to its own internal dynamics as it is buffeted between the use of labour and resources needed to live everyday life. It includes the embodied labour (paid and unpaid) in conjunction with the resources, such as those of land, 'nature', time, technology, and increasingly capital, that enable human and non-human life to occur, the emotional and material needs of everyday life to be met, as well as hopes and dreams for the future, and the material social practices that constitute the organization of daily life and life over generations to take place.[7] It is about the process of the production of value – both use and exchange value – moulded through the spatialities and temporalities of the everyday and determined through differentiation and struggle.

Social Reproduction and the Urban

The feminist political economy analyses of social reproduction discussed above, and their recognition of the need to situate processes of social reproduction – in bodies, households, institutions, and processes of globalization – has yet to extend to the urban. Reorienting social reproduction from the household to the global capitalist system at large, not least because 'the renewal of labour-power occurs in, and through, the policing of borders, flows of migrants and the remittances many send to their countries of origin, army camps, refugee camps, and other processes and institutions of a global imperialist order' (Ferguson *et al.* 2016, p. 31), social reproduction theory has tended to treat the urban merely as a spatial and empirical accoutrement. In this way, the question of space, spatiality, and spatial forms in contemporary social reproduction theory become naturalized to the phenomena under consideration. In other words, it is not that an urban spatial-blindness marks these theories; rather, urban space does not figure as an analytic category in the making of these theories.

Feminist political economy has yet to rethink social reproduction as a feminist urban problematic, namely that the urban is increasingly the site and urbanization is increasingly the process through which social reproduction takes place. Why do we argue this? Certainly we cannot ignore that the world's population is now predominantly living in places called urban (such as towns, suburbs, cities, megalopolises, and so on). And we cannot overlook that, within the next few decades, it will be approximately two-thirds of the world's population living in urban places, owing not only to rising rates of urbanization in Southern cities (through natural increase and the movement of the world's rural population into urban places), but also to the reclassification of rural areas into urban ones.[8] Our argument is driven primarily, however, by the realization that it is now urbanization, the engine of this growth and movement, that increasingly drives capitalism. Harvey's voluminous work on the urban process under capitalism and the 'secondary circuit' of capital, has shown how surplus capital is turned into fixed assets of land and real estate (i.e. the built environment). Others have pointed to the increased embedding of the state into urbanization processes. Especially in Southern cities, urban land development – through infrastructure projects, real estate for local elites, and mega projects – are often now prioritized over the provision of jobs and industrial development (Schindler 2017; Goodfellow and Owen 2018). But it is arguably Lefebvre's (2003 [1970]) thesis on urban modernity in crisis in *The Urban Revolution*, in which he theorized a trajectory of the replacement of the industrial city through a process of 'complete urbanization', that has best understood the role of the urban beyond capital accumulation and class-based struggle.

As we wrote previously (Ruddick *et al.* 2018), from the late 1940s to the early 1980s, Lefebvre followed the transformation of everyday life, to formulate a concept of the urban revolution, which he invested with two meanings. In the first, the urban revolution inverts relationships between the pre-capitalist rural and the 'urban' and subsequently the relation between capitalist industrialization and capitalist urbanization: 'The "rural" no longer produces the "urban", but the reverse. Moreover, the urban is no longer merely an effect of capitalist industrialization. Once produced, the urban does not depend on industrialization for its own continuity; it becomes capitalism's opening to different labour processes through a reorganization of socio-spatial relations' (Ruddick *et al.* 2018, p. 394). Lefebvre referred to urban society's transcendence of industrial society as the engine of capitalism, as processes of 'implosion and explosion' and of concentaration and dispersion, in which cities could be understood as zones of agglomeration that themselves implode, fragment, and destruct while also extending their infrastructural reach deep into previously remote areas (Brenner 2014).

It is in the second sense in which Lefebvre uses the urban revolution – as a shift in the site of struggle from the factory to the everyday – that he opens a space for social reproduction and the urban as a ground for the formation of difference, 'alluding to the potential for a new politics of urban revolution, which can transform everyday life in all its aspects' (Ruddick *et al.* 2018, p. 394). Beyond Lefebvre, however, rarely have (non-feminist) critical analyses of the urban turned to the relationship between urbanization and social reproduction.[9] And yet social reproduction is inexorably implicated in driving crises of capitalism (Briggs 2017). As Norton and Katz (2017, pp. 7–8) state: 'A crisis of social reproduction occurs when existing social, political economic, or environmental conditions and relations can no longer be reproduced…. Likewise, a crisis of social reproduction occurs if the labor force cannot be reproduced in a given time and place or find the means to labor productively in a given setting.' Crises of social reproduction, alongside climate and environmental crises, war, conflict, and the resultant poverty and lack of livelihoods, have resulted in the displacement of millions to and within urban places, either within their country of origin or beyond.[10] Urban life, marked by unprecedented levels of migration and inequality,[11] has led David Harvey (2014, p. 60) to note that: 'The massive forced and unforced migrations of people now taking place in the world, …will have as much if not greater significance in shaping urbanization in the 21st century as the powerful dynamic of unrestrained capital mobility and accumulation.' Not least, people on the move and the deepening of inequality from increasingly unregulated rounds of capital accumulation has loosened the relation between the state, capital, work, and labour, increasing the myriad ways in which lives are reproduced outside the wage. In the 21st century, migration, forced and unforced, is primarily a stake in a future, a stake in life itself.[12]

Following Lefebvre, we understand the urban as the conceptual knot mediating between the everyday ontological struggles of oppressed peoples, and the global spatial restructuring of hegemonic modes of production. However, rethinking the conceptual status of the urban as mediating does not confer it with an untethered epistemological salience and autonomy, thereby overriding the processes, lives, struggles, and subjectivities it is supposed to explicate. It is through social reproduction as method (as opposed to this epistemological autonomy), that the processes of urbanization, including its undoing, become 'knowable', albeit never entirely known, due to the urban's undecidability. In this way, we argue that a contemporary consideration of the spatial organization of our social lives needs to investigate the ways in which the processes of urbanization themselves are in need of explanation through social reproduction.

Whether in situations arrived at through displacement or through decades of *in-situ* neglect, the capacity for the social reproduction of everyday urban life is being eroded, characterized by uncertainty, insecurity, and disposability. The rise of precarious labour is driven in part by the desire of corporations to keep down costs – that is monetized subsidies to social reproduction: zero-hour contracts, payment below the minimum wage, short-term contracts, in short the 'gig economy', increasingly characterize the world of work, underpinned by capital's reduced commitment to the social wage and social contract. Simone's (2009) research across multiple cities reminds us that 'people as infrastructure' is not a new phenomenon; informal employment has always been an inherent part of capitalist systems of production. Precarity and insecurity, however, are now the primary material and emotional conditions through which social reproduction is instantiated, whereby the devolution of responsibilities onto the individual is not imposed but rather has become an accepted norm as it articulates with other commonsense understandings and becomes entrenched in socio-spatial practices.

The practice of migration, and its growth globally, is also partially a manifestation of the financialization of society as migration has become a way for individuals to navigate risk in the absence of the state providing conditions for their social reproduction.[13] In particular, the increasing financialization of social reproduction has influenced the ways in which urbanization takes place and is experienced. It is only slightly over a decade since the subprime mortgage crisis in the United States spread globally, generated by the restructuring of lending through the predatory pursuit of subprime mortgages, which centred on urban neighbourhoods, adding to the deepening of inequality, displacement, and austerity politics.

There were a number of pressures directly related to the financialization of social reproduction, that increased vulnerability, not the least of which was to increasingly entreat low-income Latinx and African-Americans, who had previously been redlined out of the housing market, to monetize their home-space, as a retirement plan and investment. Wade (2009, p. 40) reports that in the United States in 2008 alone 'more than 3 million houses were foreclosed in 2008, meaning that about 10 million people shifted into rented accommodation, vans or shelters' (quoted in Feldman, Menon, and Geisler 2011, p. 12).

In the face of such devastation, we turn to the chapters in this volume to explore the social reproduction of everyday urban life. Building on feminist urban theories and social reproduction feminisms, the chapters shed light on different aspects of the relationship between the urban and social reproduction, within different contexts but always through socio-political action. In what follows, we outline how the book's contributors address not only this relationship but also their irrevocable relation to questions of urban feminist knowledge production. We recognize themes that speak directly both to the production of the urban in relation to infrastructures, labours, and subjectivities, and the politics of this production, which engage the challenges of decolonizing feminist urban knowledge production and methodologies.

Making the Urban Through Feminist Knowledge Production

Infrastructures

The ethos of liberal citizenship in Western democracies finds one of its most crisp articulations in the presuppositions frequently relied upon regarding everyday urban life. As a benchmark of modernization, urban forms present people with 'proper' infrastructures through which an individual's life-chances in the capitalist market prosper, thereby ensuring a 'successful' integration into the public life of civil society. However, as is now abundantly clear, the relationship between capitalism, modernity, urban forms, and the reproduction of people's everyday lives is not as straightforward as this modernist narrative suggests. Even for those historical instances in the global North in which there is a resemblance to this narrative, feminist and postcolonial scholars demonstrate that it is invariably subtended by gendered, racialized, classed, and sexualized operations of power. Increasing neoliberalization, austerity, and precaritization, both in the global South and North, has been creating other everyday lives for the majority of urban residents, for which no blueprint is available; neither infrastructure nor people's access to it can be taken for granted. A number of the chapters in this volume collectively argue that it is the intrinsically agentic nature of the social reproductive work of those pushed into precarity that mediates between infrastructures and the urban, highlighting the centrality of social reproductive work in the making of the urban.

Before turning to these contributions, we briefly consider Mbembe's conceptualization of 'superfluity' and Simone's conceptualization of 'people as infrastructure' in order both to interrupt hegemonic ontologies of the urban and to situate the contribution of these chapters in an ontologically reflexive context of knowledge production. The work of Mbembe and Simone show us the limits of metropole capitalism's teleological social ontology, reminding us how the social ontology of the urban of former colonies is formed differently and how, within this latter social ontology of the urban, people become infrastructure (see also Roy 2009).

In considering the spatialization of an African metropolitan modernity as an historically specific urban form, Achille Mbembe offers the concept of 'superfluity', referring to both 'the dialectics of indispensability and expendability of both labour and life, people and things' and 'the obfuscation of any exchange or use value that labor might have, and to the emptying of any meaning that might be attached to the act of measurement or quantification itself insofar as numerical representation is as much a fact as it is a form of fantasy' (Mbembe 2004, pp. 374–375). In this way, superfluity can facilitate a socio-spatial investigation within the interstices of political, economic, biopolitical, and psychic approaches to the urban. Drawing on Simmel, Mbembe argues that 'the ultimate form of superfluity is the one that derives from the transitoriness of things' (Mbembe 2004, p. 399).

This transitory character of urban life constitutes the omitted predicate of a colonial urbanity that represents itself through the fixity of its infrastructure, as the immutable monument to its historical teleology, to its 'developed' and 'civilized' telos. AbdouMaliq Simone's landmark conceptualization of 'people as infrastructure' in African cities provides a deeper understanding of this transitoriness by turning to the 'incessantly flexible, mobile, and provisional intersections of residents that operate without clearly delineated notions of how the city is to be inhabited and used' (Simone 2004, p. 407). People as infrastructure illuminates the provisional and precarious workings of the 'the dialectics of indispensability and expendability of both labor and life, people and things' (Mbembe 2004, p. 374) in relation to the reproduction of the urban by indicating 'residents' needs to generate concrete acts and contexts of social collaboration inscribed with multiple identities rather than in overseeing and enforcing modulated transactions among discrete population groups' (Simone 2004, p. 419).

Focusing on expressive urban cultural practices in the wake of 'natural' disasters in Haiti and Puerto Rico, Nathalia Santos Ocasio and Beverley Mullings (Chapter 2) examine the conditions of possibility of people as infrastructure through a generative theoretical conversation between social reproduction, Simmel, and Simone. They ask their readers to consider how a society is possible in disaster- and debt-stricken contexts of austerity capitalism when the urban infrastructures of everyday life are devastated. In the course of their analytical deliberation, they first turn to Simmel's conceptualization of forms of sociation as the unceasing emergences and interactions that produce the unity of society within which its members live. One such form of sociability, according to Santos Ocasio and Mullings, might be found in expressive cultural practices, in particular music. Performed as a part of social reproductive labour that ensures, amongst other things, the reproduction of intergenerational linkages between the Caribbean and its African inheritances, expressive cultural practices provide the conditions of possibility of people as infrastructure by making sure sociability itself is imaginable and enacted in the aftermath of disasters. Santos Ocasio and Mullings evoke the importance of deeply ancestral forms of music, dance, and gathering, in the form of intergenerational memory and knowledge sharing practices. Therefore, as opposed to taking social reproduction as the work that makes all other work possible, they point to a series of practices of social reproduction which are not tethered to the economic but express their own logics, drives, and histories, therefore turning their attention to sociation as a zero point of sociability.

In Chapter 8, Natasha Aruri also investigates the destruction of the social and its spatial preconditions and effects. In her chapter she traces 'socio-cide' in Ramallah, resulting from ongoing colonial violence and espoused through neoliberalism, via the urban development of its built environment and infrastructure. For Aruri, social reproduction, broadly construed, becomes an analytic not only to which city- and place-making are re-oriented but also through which decolonization might be imagined in a context where Palestinian lives are rendered superfluous. Demonstrating how individual parcelization and zoning upon which urban development in Ramallah is premised are reproductive of a colonial logic of governance and unusable antispaces, Aruri argues for a common land ownership model that would enable reconfiguring urban land as a continuum that is open to connections and relationalities. In this way, antispaces could be re-socialized as spaces of social reproduction through which decolonial resilience and resistance could be collectively organized against the colonial dispossession and occupation of Palestinian land. Such centering of social reproduction in the production of space better prepares the city to deal with uncertain and transitory vital infrastructures, such as the water infrastructures in Ramallah, which are the customary targets of colonial military violence. Understanding the urban as a commons oriented towards social reproduction, according to Aruri, helps dissipate colonial assumptions about the stability of life and vital infrastructures.

Thinking of people as infrastructure in conjunction with social reproduction is the theoretical focus of James Angel (Chapter 5). He draws on Ruddick *et al*'s. (2018) imperative of orienting analytical attention to the social ontology of the urban, lest we run the danger of forgetting people, struggle, difference, and history in our accounts of the production of the urban, ending with an autonomous epistemological category of the urban. Contributing to a social ontology of the urban that is centred around praxis, Angel focuses on the Catalan activist network la Alianza Contra la Pobreza Energética (the Alliance Against Energy Poverty, APE). He demonstrates that APE's feminist urban praxis is 'premised upon the creation of more caring and collectivized modalities of social reproduction' for those who do not have access to vital infrastructures, such as gas, electricity, and water, for their survival. Angel's analysis illuminates how social reproductive labour and people as infrastructure become intimately entangled during the processes of reproducing the urban and life within it, thereby providing us with the ethnographic details of a social ontology of the urban in Catalonia.

Tom Gillespie and Kate Hardy contribution (Chapter 11) also draws upon Ruddick *et al*.'s conceptualization of the social ontology of the urban. Operating within a framework of feminist comparative urbanism, Gillespie and Hardy discuss two urban social movements: The Asociación de Mujeres Meretrices de Argentina (AMMAR), a sex workers' union in Córdoba, Argentina, and Focus E15, a housing campaign in London, UK. Focusing on how these all-women or women-led grassroots activist organizations both reproduce the urban and their own subjectivities within the process, Gillespie and Hardy's comparative methodology is hinged upon what Ruddick *et al*. call 'infrastructures of social reproduction' (Ruddick *et al*. 2018, p. 396). Gillespie and Hardy employ ethnographic detail to analyse these movements' struggles over the infrastructures of social reproduction – of health, education, and housing – and show how they employ various tactics, such as 'demand[ing] access to existing infrastructures, creating autonomous infrastructures, and co-producing new infrastructures with the state' (Gillespie and Hardy, Chapter 11). Their chapter not only highlights urbanization as 'an open process determined through praxis, by actual people making the world they inhabit' (Ruddick *et al*. 2018, p. 399) but also signals the importance of historical difference in the constitution of the urban and to the fact that despite the seeming universality of social reproduction and its infrastructures they are always marked by this difference.

Meera Karunananthan (Chapter 7) also focuses on struggles over the infrastructures of social reproduction, through an account of the feminist network Solidaritas Perempuan's campaign for the right to water in Jakarta, Indonesia. Through an intersectional feminist approach, Karunananthan examines the ways in which human rights discourse might be employed to make visible urban poor women's social reproductive struggle with privatized drinking water systems. Moreover, Karunananthan elaborates on how the right to water activism might help to recast the Trotskyite transitional programme in a feminist manner to recuperate subaltern women's revolutionary subjectivity and expertise, which, she argues, is often unnoticed by the male leadership in established leftist groups. Karunananthan demonstrates that through the demands of collectivizing social reproduction in relation to urban water infrastructures, feminist activists of Solidaritas Perempuan Jakarta (SPJ) give priority to use-value production at the expense of exchange-value production, thereby reversing transnational capitalist logic and exposing its gendered violence at the urban, household, and bodily scales. For her, the sites of social reproduction and social reproductive labour are crucial in defeating capitalism in cities of the global South, and a social imaginary of a 'just city' becomes tenable with the reclamation of 'the labour power of women whose unpaid work has served to subsidize' postcolonial capitalism.

Subjectivities

In Ruddick *et al.* (2018), we approached theories of the urban through a primary focus on social ontology, aiming not simply to situate the subject as an intellectual problem – 'who acts' – but to centralize processes of subjectivation. In this volume the authors track the empirical folding and unfolding of subject formation in ways that show the subject is not a mere artefact which can later be 'situated' within more or less determined processes of urbanization. As 'the sedimented outcome of material social practice' (Mitchell, Marston, and Katz 2004, p. 10), subjects are not only constituted relationally, but also by their physical environments which play a role in constituting intersubjective encounters. Beyond urban form, Hoffman's (2014) research on how the politics of urban governance in Chinese cities leads to the normalization of the self as cosmopolitan – however incomplete – further evidences how subjectivities are shaped through urbanization, producing subjectivities that are not only *in* but also *of* the urban. She shows how urban politics is located in the constitution of new categories of subjects in Chinese cities giving rise to new modes of self-governance such as self-enterprise, volunteerism, and charitable giving. In the Chinese context,

'This has produced particular kinds of cities (entrepreneurial, financially "efficient") emerging in tandem with particular kinds of subjects (professionals and volunteers)' (Hoffman 2014, p. 1583).

The formation of the self leads to a range of political possibilities – some collectively revolutionary, others highly individualistic. The contributors in this volume are not searching for a new or singular revolutionary subject, one which will indicate the exit, complete or otherwise, from any capitalist mode of production. Neither do they tend to an over-presence of the urban subject as a replacement for the industrial worker as the collective agent of revolution. As Mantha Katsikana describes in Chapter 4, the barriers to the construction of an anarchist and anti-authoritarian commons in Athens, many revolutionary movements continue to emphasize 'the accumulation and display of male power' as opposed to the 'affective and connective labour practices' needed for the 'social and emotional change necessary to build and reproduce durable relationships'. Katsikana's work speaks to the need to demasculinize radical and revolutionary subjectivities in order to better understand and appreciate how the 'emotional needs and manual tasks necessary for the everyday context of collective actions' are undertaken primarily by women in these movements (as well as the broader context in which many contemporary transformative movements are led by queer women and women of colour). Her work points to the collective renegotiations necessary to enable the malleability of subjectivity as a relational form of collective self-understanding. Katsikana's study is a welcome antidote to the now well-travelled theories of revolutionary urbanization and revolutionary subjectivity within urban studies that have overlooked the role of social reproduction and the fabric and texture of everyday life in promulgating transformations. Seeing greater potential for an engagement with gendered subjectivities in the shift from the factory as the heart of revolutionary struggle, feminist scholars have argued that struggles are not just about belonging in the city but also about how the city belongs to those whose invisibilized and unpaid labour maintains the urban (Buckley and Strauss 2016). We also see this engagement in Karunananthan's work with SPJ, which 'calls for Marxist debates regarding revolutionary praxis to be re-examined in light of both the constraints faced by women living in the margins of cities of the global South as well as their aspirations.'

What is evident from the research reported in these chapters is that for increasing numbers of people their own social reproduction is increasingly precarious and provisional, falling outside of and challenging the norms of a neo-liberal political subjectivity. Precarity, generally, speaks to the disintegration of stable societal bonds, social protections, and senses of entitlement and belonging, creating lives structured by insecurity, eroding the possibility of life itself (Puar 2012). Precarity is not born of the economic project of neoliberalism, but is a signifying characteristic of it. With increasing inequality, the reorganization of economic and social relations in the context of the hollowing out of social and political institutions, and in the absence of infrastructures (Butler 2012), we can think of the governance of precarity as designating not only working and living conditions but also subjectivities and embodiment, and therefore agency. And as such, as Lauren Berlant (2011) notes, precarity is also a structure of affect internally inculcated into subjectivity via anxiety as the dominant lived experience of insecurity. From the point of view of organizing the everyday, the increasing inability to replenish the self, materially and mentally, has had enormously deleterious effects. Decades of neoliberal scouring out of the social and the permeation of the values and organizing principles of finance capital into society more broadly are leading to new forms of subjectivation, with a depleted, indebted, and anxious subject now prevailing across a wide variety of places.

It is within this context that Simone's notion of 'people as infrastructure' and Caroline Moser's notion of women's triple role – of reproductive, productive, and community-managing activities – come together to highlight how a strongly gendered division of labour not only in the household but also within communities, underlies, 'economic collaboration among residents seemingly marginalized from and immiserated by urban life' (Simone 2004, p. 407). Belinda Dodson and Liam Riley (Chapter 10) illustrate this convergence with reference to food systems in three African cities: Kitwe in Zambia, Kisumu in Kenya, and Epworth in Zimbabwe. Within households in these cities women are largely responsible for food procurement, allocation, and preparation, and in the broader urban food economy they are 'important actors ... as traders, processors and producers, especially in the urban informal sector.' It is their paid and unpaid time engaged in food-related labour that helps reproduce patriarchal family structures and limits women's participation in other activities, placing strictures on their subjectivity formation. This particular 'mode of provisioning and articulation' speaks not necessarily to an 'efficient deployment' of the 'energies of individuals' (Simone 2004, p. 407) but to 'the gendered social forms and practices that reproduce life, family and labour in conditions of urban precarity' (Dodson and Riley).

With the urban as the primary mode through which capitalism endeavours to organize the social, political, and economic realms, a number of other chapters also highlight how processes of subject formation in the realm of social reproduction reveal fractures in capital's attempts to address those relations which capitalism has not yet been able to fully undo, incorporate, or defeat. For example, Angel argues that it is 'through engagements with irregular infrastructural connections, that new ways of navigating and producing the city (and urban subjectivities) are being performed.' In other words, APE activists saw themselves as engaging in a feminist praxis, given their efforts to 'sustain life' in 'collective and egalitarian ways'. Although they show glimpses of a more emancipatory urban future, these struggles were also replete with the violence and precarity that leads to the formation of fragmented fugitive and indebted subjects, who had to resort to illegal occupation and illegal connections to the city's formal water and electricity networks. And while APE's collective actions saw success in legalizing water connections, they were less successful in securing legality for electricity connections, leaving poor inhabitants in a state of insecurity.

Also with a focus on the urban politics of infrastructure, Gillespie and Hardy's account, in Chapter 11, of their comparative study of AMMAR and Focus 15 shows how 'women's subjectivities' that emerged from these engagements changed over time 'from victimised, stigmatised and invisibilized subjects to agential actors with collective strategies for changing the conditions in which they live.' Faced with considerable ignominy, as sex workers and single mothers, both groups 'initially mobilised around notions of motherhood.' Despite the positive narratives of political motherhood that arose in the 1970s and 1980s in Argentina with the Madres de la Plaza de Mayo, the sex workers' engagement with union activities eventually led to them identifying primarily as members of the working class. In both Córdoba and London, women's changing subjectivities led to new demands. In the UK, the single mothers also moved on from an identification solely as mothers to being housing rights campaigners, as they increasingly came to recognize that they were part of 'a much wider housing crisis that had not only gendered, but also classed and raced dimensions.' In Cordoba, the focus shifted from police repression to demand for access to infrastructures of social reproduction in education and healthcare.

Santos Ocasio and Mullings (Chapter 2) address the development of a collective subjectivity in a context of the absence of any reference points for an imagined future. In their exploration of the role of musical expressive practice in urban social reproduction, they address the 'impulse to sociability' – what brings people together and how that coming together sustains and enables the intergenerational and ancestral (re)connection to a sense of collective subjectivity, belonging, and liberation. Integral to sustaining a collective subjectivity is not only the unpaid work involved in social reproduction, including the passing on of knowledge, social values, and cultural practices, but also the forms of sociability and collective critique found in the expressive labour of making music.

Decolonizing Feminist Urban Knowledge

Our broad project of advancing a feminist urban theory for our time is predicated on recognizing the need to decolonize feminist knowledge production about the urban, including within this book. Coloniality, or the patterns of power resulting from colonialism that have shaped subjectivities, political and economic power, and knowledge (Maldonado-Torres 2007; Noxolo 2017), brings into view the way in which historical structures of gendered oppression, such as patriarchy and heteronormativity, work in concert with structures of class and racial ordering to shape contemporary urbanization. While postcolonial theory has long analysed how colonial power has shaped knowledge and global systems of economic, political, and cultural ordering emanating from Eurocentric epistemologies, decolonial theory from Latin American and Caribbean perspectives has theorized the relationship between coloniality and modernity, and liberation from coloniality as a political project. Latin American feminist traditions have further sought to critically interrogate decolonial scholarship through a '*descolonial* approach' (Esguerra Muelle, Ojeda, and Fleischer, Chapter 9), emphasizing the role of gender oppression in colonial power and the need to connect with ongoing anticolonial movements in Latin America and the Caribbean. At the same time, Indigenous scholars have sought to move beyond postcolonial concerns with representation to emphasize the lived voices and experiences of colonized subjects, particularly in spaces occupied by settler-colonists where Indigenous peoples and Indigenous geographies continue to be subjected to processes of dispossession (de Leeuw and Hunt 2018).

Urban theorists have used postcolonial analysis to argue for the heterogenization of urban theory, particularly through greater attention to cities in the global South (McFarlane 2010; Robinson 2011), and through calls to take historical difference seriously as a constitutive element of the urban (Roy 2016a; Jazeel 2019) – issues which assume particular salience in the context of the ascendance of comparison as a mode of analysis in urban studies in the 21st century (Nijman 2007; Ward 2010). Amongst the varied urban geographies presented in this volume are several chapters that provide insight into the formation of the urban through the predatory relations put in place by 'economies of dispossession... those multiple and intertwined genealogies of racialized property subjection, and expropriation through which capitalism and colonialism take shape historically and change over time' (Byrd *et al.* 2018, p. 2). In particular, they shed light on urban formations in the contemporary phase of capitalism, its logics of speculation, expropriation and dispossessive financialization, and the related biopolitical and necropolitical regimes of racialized value that they inaugurate (Tadiar 2013; Hong 2018).

In Chapter 3, Emily Fedoruk's analysis of a public mural quoting the words of Qayqayt First Nations Chief, Rhonda Larrabee, in the Vancouver suburb of New Westminster, British Columbia, highlights Indigenous social reproduction amidst ongoing processes of colonial dispossession in settler colonial Canada. Fedoruk situates the social reproduction of the Qayqayt First Nations into the broader context of settler-colonialism, thereby avoiding collapsing the ongoing violence of capitalist settler-colonialism into the violence of contemporary urban capitalism, making it possible to reflect on these different forms of violence relationally and historically. More importantly, by using social reproduction in this methodological way, this chapter directs us beyond social reproduction, towards Indigenous ontologies of life and history. In their exploration of the legacies of plantation economies and neoliberal urban transformation in the Caribbean, Santos Ocasio and Mullings (Chapter 2) discuss the ways in which processes of 'disaster capitalism' (Klein 2007) and 'debt imperialism' (Kim 2018) shape urban dispossession in the wake of environmental disaster in Haiti and Puerto Rico. They commit to what Frantz Fanon (1961, p. 210) terms 'passionate research', in order to seemingly recover 'beyond the misery of today, beyond self-contempt, resignation and abjuration, to some very beautiful and splendid era whose existence' provides communities battered by disaster capitalism with the tools to rehabilitate themselves and others. In her chapter on spatial politics in Ramallah, Natasha Aruri (Chapter 8) discusses the confluence of neocolonialism and neoliberal modes of urban development in a context of ongoing militarized settler colonial occupation in Palestine. As a primary vehicle through which finance capital feeds into contemporary urbanization, Aruri tracks the proliferation of speculative capital in the real estate market and its deleterious impact on possibilities for everyday social reproduction in Ramallah.

Providing insights into the ways in which the urban is produced through the distribution of social reproductive labour across transnational circuits of care and labour migration, Chapter 6 by Faranak Miraftab and Chapter 9 by Esguerra Muelle, Ojeda and Fleischer demonstrate that at a transnational scale processes of social reproduction are organized through the legacies of historical colonial relationships, as well as racial divisions of labour in contemporary imperial formations. Miraftab analyses the transnational circuits of social reproduction that come to serve crises of capitalism in this latest era of global capitalism, enabling, amongst others, the revitalization of the United States 'rustbelt' town of Beardstown, Illinois. She explores the global restructuring of social reproduction, through the place-making practices of migrant workers from Central America and West Africa, and how social reproduction work is made invisible not only through its gendered normalization but also through its spatial fragmentation, both across the globe and within existing postcolonial racialized urban hierarchies. In doing so she challenges the racialization and criminalization of these migrant populations in nationalist discourse to render visible the transnational contributions of their labour. Esguerre Muelle, Ojeda, and Fleischer explore the multiple forms of violence that connect internal displacement in post-conflict Colombia, resulting from war and rural dispossession, with the re-enactment of colonial gendered and racialized labour relations in transnational care migration networks between cities in Colombia and Spain. Through a collaborative multi-sited ethnography conducted in four Colombian cities – Cali, Cartagena, Bogotá, and Medellín – and two Spanish cities – Madrid and Barcelona – they explore how Columbian women in Spain become trapped in a cycle of migration-return, effectively disposed to sustain uneven processes of urban production and, how in Colombian cities, *madres comunitarias* (communitarian mothers) conduct a form of underpaid care work sustained mostly by women of rural origin who have been forcedly displaced. Their work shows how the intertwined dynamics of war and globalized capital have forged a problematic geography of urban-based care work through which colonial power is constantly re-enacted.

The task of imagining a feminist urban theory that is capable of both analysing these recursive colonial logics and of envisioning possibilities for decolonization returns us to the political conjuncture of the epistemological, the methodological, and the ontological at the core of feminist philosophy. Given the deep imbrication of knowledge systems in the proliferation of colonial power, decolonization necessitates an interrogation of knowledge creation processes in terms of who generates theory, how, and the ends that theory serves (Jazeel 2019). The creation of possibilities for decolonization within the academy through privileging the 'singularity of indigenous, southern and subaltern narratives' (Jazeel 2019, p. 11) is contingent on meaningful attempts to pluralize and heterogenize the bodies and voices that constitute the epistemic communities of the academy. As Tuck and Yang (2012) have reminded us, 'Decolonization is not a metaphor.' Rather, it is a radical and transformative political practice that belongs outside the confines of the academy. In this context, the decolonization of knowledge frameworks within and beyond the academy serves as an aid to political efforts to end colonial domination, from the dismantling of racist epistemological frameworks that underpin Eurocentric power, to Indigenous campaigns for the radical restructuring of relationships to land, resources, and the environment (Esson *et al.* 2017). As Indigenous scholars have long argued, the impulse to render legible and explicable, which is inherent in intellectual cultures of subsumption, may militate against the ontological possibilities proliferating from attempts to reach for a decolonial horizon (Hunt 2014; see also Santos Ocasio and

Mullings, Chapter 2 and Fedoruk, Chapter 3). May we go so far as to ask whether an analysis of the creative and insurrectionary energies of decolonial praxis requires that we question and disinvest from the framework of social reproduction?

In the ongoing quest for locating 'new geographies of theory' in urban studies (Roy 2009), for example, Jazeel's recent call for a focus on 'singularity' as a way to open up to difference in knowledge production provides a useful epistemological intervention that begins by rendering visible disciplinary cultures of subsumption, which serve to reduce 'examples and cases to exchangeable instances, or conceptual givens, for the benefit of a disciplinary theory culture located in the EuroAmerican heartland' (2019, p. 11). If we were to privilege singularity, we may have to contemplate that decolonization as praxis may fall outside of any one overarching explanatory framework, including that of social reproduction, and may indeed exceed our known epistemological grids of representation (Jazeel 2019). For example, Santos Ocasio and Mullings' chapter on the role of expressive musical practices in enabling the reconstruction of relational community infrastructures in the event of natural disaster, and in asserting critiques of ongoing imperial and colonial dispossession, offers a compelling example of urban praxes that manifest 'affective and grounded alternatives to economies of dispossession' (Byrd *et al.* 2018, pp. 11–12). Santos Ocasio and Mullings conclude their analysis by casting doubt on the transformative potential of the expressive arts to effect material change in the world. It might be worth asking could we gain more in dwelling in the space of the *unspeakable* evoked by the Haitian song leader they cite in their article, who says: 'If you don't have this reaction instilled in you, you cannot understand it; it's inexplicable!'

Following 'fragments', translation and untranslatability, and poetics, are amongst the tactics put forward by Jazeel (2019) for working towards singularity. In Chapter 3, Emily Fedoruk traces the poetics of urban space through fragments of text and in so doing reflects on the role of illegibility in rejecting settler colonial regimes of recognition of Indigenous people in Canada. Juxtaposing it with another poem that also appears on the same building (by architect Graham McGarva), which adopts a colonial voice, she articulates some of the complexities of authorship. Ruminating on the space between translation and untranslatability, the written and unwritten, Fedoruk examines the potential of a fragmented poem in a public space to reclaim the survival of Indigenous people against the genocidal processes of colonial place-making. Fragments are also present in Chapter 8, with Natasha Aruri's call to reclaim the 'antispaces' resulting from colonial logics of spatial dissection in Ramallah, and to re-imagine the possibilities of these forgotten spaces for grounding a politics of communal regeneration and, ultimately, decolonization in a context of ongoing military occupation. Such readings of fragments and untranslatable utterances map 'decolonial geographies as constellations in formation' (Daigle and Ramirez 2019, pp. 79–80), which evince tactics of refusing and resisting racialized economies of containment, displacement, and interconnected violences against lands, spaces, and bodies.

Methodologies

Positionality and reflexivity have been key methodological strategies in feminist scholarship since the mid-1980s, foregrounding the unequal power geometries of knowledge production (Harding 1986; Haraway 1988; Mohanty 1988; England 1994; Kobayashi 1994; Nagar and Ali 2003; Peake 2016). In keeping with this long-standing feminist practice of recognizing that all knowledge is situated in particular places, we asked contributors to this volume to reflexively locate themselves in relation to their work by explicitly addressing their positionality. There was considerable variability to the ways in which authors responded to this invitation, reflecting the multiple geographies they were situated in, and multiple vectors of power that are mapped by the transnational research networks evoked in this volume. The contributors have highlighted that positionality is not a straightforward matter; scholars may occupy complex and multi-layered positions drawn from personal biographies of mobility, migration, or displacement, which cast them simultaneously as settler colonial subjects, as diasporic and transnational subjects, and as both 'insiders' and 'outsiders', with experiential or empathetic connections with their research sites and subjects (Santos Ocasio and Mullings, Chapter 2; Miraftab, Chapter 6; and Aruri, Chapter 8). However, as Indigenous and feminist scholars have argued, reflexivity is about political accountability to the people and places one is working with (Nagar 2002). Esson *et al.* (2017), for example, claim genuine decolonization requires the cultivation of critical consciousness to work in concert with activism. Several authors in this collection have situated their work in the context of participation in, and ongoing relationships with, activist communities and have illustrated how research processes are also constitutive of researchers' subjectivities (Katsikana, Chapter 4; Angel, Chapter 5; Karunananthan, Chapter 7; Gillespie and Hardy, Chapter 11).

Accounting for positionality also requires an acknowledgement of the ways in which scholars are themselves imbricated in structures of coloniality and, thus, often ambiguously placed in relation to projects of decolonization (Dodson and Riley, Chapter 10). A decolonial agenda requires a confrontation with structures of white supremacy, privilege, and racism (Esson *et al.* 2017) and its connections with ongoing econo-mies of extraction in unequal geographies of social reproduction. To this end, in Chapter 3, Emily Fedoruk's hermeneutic approach causes her to reflect on her position as a 'settler-reader' of the quote from Indigenous Chief, Rhonda Larrabee, of the Qayqayt First Nation, part of a public art work 'on unceded territories of Musqueam, Qayqayt, Tsleil-Waututh, Skxwú7mesh, Katzie, and Kwantlen Nations' in New Westminster. Her reading is, as she puts it, 'conditioned by my experiences as a white set-tler living for 25 years on Coast Salish territory'. She returns to this po-sitionality at various points in her text to forestall a possessive reading of Larrabee's text, to recognize that her reading of the text is itself tied to her own social reproduction as a knowledge producer in the academy, and to remind herself of the limitations of her own readings of Larrabee's text or even of the Indigenous feminist scholars she cites in her chapter. Fedoruk's analysis of her positionality reflects the ongoing ways in which academic research relationships, whether in urban studies or other fields, are immersed in the extractive logics that have historically structured the processes of racial capitalism and colonialism that continue to undergird economies of dispossession (Nagar 2008; Byrd *et al.* 2018). Furthermore, as Esson *et al.* (2017) have cogently argued, the deployment of discourses of decolonization within the academy is mired in a racial politics of gate-keeping and instrumentalization, wherein the use of decolonial language by non-Indigenous and white academics serves to reproduce coloniality by galvanizing the very structures of white supremacy that reinstate white privilege (see also Duarte and Belarede-Lewis 2015; Noxolo 2017; de Leeuw and Hunt 2018).

Feminist urban theory must be capable of critically engaging with these persistent historical and political realities if it is to avoid colluding with a politics of co-option, disempowerment, and reinstatement of racial (and particularly white) privilege and serve as a transformative tool for enact-ing decolonization. Reflexive analyses of positionality have gone some way in addressing these realities; as methodological strategies they un-derscore the need to remain continually vigilant to enduring erasures and new occlusions that might be constituted, even as the ethics and politics of research, representation, reflexivity, reciprocity, responsibility, and soli-darity are being attended to in ever more nuanced ways through the work of scholars who elaborate feminist, postcolonial, decolonial, and intersec-tional approaches to knowledge production and praxis (Faria and Mollett 2014, 2018; Daigle 2019; Nagar 2019).

The diverse research designs that contributors to this volume have deployed also highlight how they grapple with these methodologi-cal dilemmas of doing research as they seek to produce non-totalizing narratives of the urban. They fall into three (not mutually exclusive) clusters of: non-extractive praxis-oriented research; relational multi-sited research; and research based on a use of mixed methods.

The contributions by Katsikana (Chapter 4), Angel (Chapter 5), Karunanthan (Chapter 7), and Gillespie and Hardy (Chapter 11) favour 'non-extractive' collective feminist praxis to generate knowledge that can 'resource' struggles and be useful to movement actors. In pursuit of this goal, Angel navigates through the responsibility of his dual identity as a scholar and activist, and ultimately 'resources' the struggles he engages in by drawing upon his bilingual skills to translate movement literature and by seeking to build solidarity between activists located in the UK and Spain, such that these activist groups can reinforce and lend support to each other. For their part, Gillespie and Hardy elaborate a 'dialogic collaboration' method, which grants epistemic privilege to movement actors and deploys comparison to design research that, through ongoing dialogue, asks research questions that are relevant to movement actors, thereby 'co/produc[ing] knowledges that "speak" the theoretical and political languages of communities' (Ali and Nagar 2003, p. 365). Karunanathan, too, embodies a scholar activist praxis as she seeks to resource Solidaritas Perempuan Jakarta, by amplifying their local struggle to the international media, standing with them as an ally to highlight their role as knowledge producers. Finally, Mantha Katsikana (Chapter 4) addresses persistent contradictions and conflicts arising in Greek anti-authoritarian movements, spaces, and struggles in which she actively participated, directing the reader's attention to the everyday praxis of the 'personal is political', especially as it shapes an urban commons that is all too often figured as implicitly, if not exclusively, masculine.

A further set of approaches, broadly encompassing comparative, relational and multi-sited, are at work in the chapters by Miraftab (Chapter 6), Muelle, Ojeda, and Fleischer (Chapter 9), and Gillespie and Hardy (Chapter 11). Such relational methods are important to knowledge production in urban studies; beginning from multiple places and tracing the relational trajectories of the evolution of places is to displace the epistemic primacy that has been given to the global North, while 'rejecting any notion of pre-given "cases" or variants of a presumed universal/general process' (Hart 2018, p. 373). In Chapter 11, Gillespie and Hardy embrace 'dialogic collaboration' to link and think through their participation in a sex worker union campaign in Córdoba and a single-mother housing campaign in London. They weave elements of feminist standpoint theory, social ontology, and activist/participatory methodologies together, both to reflect on their movement-centric and historically differentiated collaborations and to create explicit linkages and dialogues between and amongst contexts that might otherwise diverge under the weight of facile distinctions between global North and South. In Chapter 6, Miraftab's relational approach introduces multiple temporal and spatial standpoints – as opposed to the single axis of a here-and-now approach that is common in social reproduction theories – to analyse the post-colonial racialized capitalist global hierarchies between the global South and North. Anchoring her research in the specific location of the US rustbelt city, Beardstown, Miraftab seeks to theorize the global restructuring of social reproduction through flows of migrants between Mexico and Togo. Similarly, Esguerre Muelle, Ojeda and Fleischer (Chapter 9) undertake a decade-long, relational multi-sited collaborative research project between South American and Spanish cities to delineate uneven geographies of care access and provision.

Unsurprisingly, and most commonly, a mix of traditional social science qualitative methods are employed by the contributors. In Chapter 8, Aruri deploys mixed methods in novel ways. In order to critically analyse the real estate development in the city of Ramallah she deploys such standard methods as semi-structured interviews, focus groups, discourse analysis of legal documents, and commentary on social media. But crucially, building upon her training as an architect, she combines these with a visual method that pays particular attention to the architectural and morphological elements of Ramallah. The value of this combination of methods not only enables her to demonstrate the importance of public space to social reproduction but also allows her to offer suggestions that have the potential to expand the imagination of Ramallite designers, planners, and spatial entrepreneurs to build 'antispaces' that reconfigure public space in such a way that new orders and modes of decolonial social reproduction can be achieved. In Chapter 4, Katsikana, draws upon interviews, participant observation, and content analysis as well as her own personal experience, in order to understand how the affective and collective labour of resistance within anti-authoritarian/anarchist movements contributes to social reproduction in Athens. While the contributors to this volume, like many other critical urban researchers, largely favour such qualitative methods, there are also those that employ quantitative methods; to produce partial, situated knowledges does not imply that qualitative methods are always privileged over quantitative methods, as methods themselves are not *ipso facto* feminist (Lawson 1995; Peake 2015). Dodson and Riley (Chapter 20), for example, deploy the data generated from quantitative surveys interpolated with those gathered from qualitative interviews to highlight the gendered nature of both the urban food system and urban food poverty in the three African cities where they work. By mixing quantitative and qualitative methods, Dodson and Riley point to the generative capacity of mixed methods feminist urban research.

The Limits of Social Reproduction

While social reproduction helps us generate deepened analyses of urbanization processes, the formation of the urban, and the lived struggles of urban residents, we recognize, that like all concepts, it has limits, including those we already discussed in relation to the imperatives of decolonizing feminist urban knowledge. As with all attempts to make theoretical sense of worlds in transformation it is wise to be circumspect about the uses of the conceptual frameworks we nurture and to both acknowledge and set our sights beyond their limits. We argue that feminist theory needs to reflect on the limitations of its main concepts and its processes and politics of knowledge production.

One such limitation concerns the collapse of social reproduction into social ontology. Social reproduction eschews the question of social ontology by presenting itself as life-making, the problem being that life is a metaphysical concept, even though it is the most material everyday experience that we all go through. In our view, it is a mistake to think of social reproduction as the production of life itself because such an approach to social reproduction can envelop every possible subject and their everyday struggles into a unitary vision. In this level of universality, it can easily be argued that all people have the same problems ('we are all in this together') and require the same solutions. However, this is to push the problem of life not only out of social ontology, but also out of history. In such a scenario, social reproduction is epistemologically operationalized as a false universal appealing to a transhistorical and transgeographical 'human nature'. Social reproduction then ceases to be a method of investigating and sustaining social ontologies, instead replacing social ontology with itself as life, foreclosing how social reproduction is taken up under different life forms within history.

Amongst the potential pitfalls of such an approach, of making social reproduction a stand-in for life itself, is that of foreclosing appreciation for and engagement with approaches that have deeply rooted political, cosmological, and ontological understandings of and orientations towards the relations that must be sustained in order that life – and not only human life – can thrive. Collapsing these relations into the relations of social reproduction in an anticipation of a unitary framework of analysis stymies the possibilities of reflecting relationally and historically on different forms of violence that render life unliveable for those who are on the receiving end of these violences, as well as for those who directly or indirectly benefit from them.

Moreover, such an approach to social reproduction runs the danger of rendering the social of social reproduction into an index of empirical varieties of oppression, as opposed to having a formational view of it whereby it is not taken for granted but understood in the historicity of its contingent precarity. For instance, slavery, ongoing settler-colonization, and the violence of contemporary capitalism and heteropatriarchy make social reproductive work extremely difficult, for in the aftermath of and during these violent regimes, the social itself needs to be reconstituted. Therefore, as opposed to taking social reproduction as the work that makes all other work possible, an approach that examines what undergirds social reproduction by focusing on the collective re-constitution of the social becomes necessary.

It follows that we need to interrogate the limits of what is signified by the 'social' in social reproduction, for example, by probing the anthropocentric conceptualization of social reproduction. The divisions within and between the human and nonhuman that underpin capitalist urbanization (Ruddick 2015) also function to ground an anthropocentrism within frameworks of social reproduction (Andrucki *et al.* 2018). The following chapters are haunted by organic and inorganic materialities beyond the human, such as water, crops, landscapes, buildings, which, however, largely come into articulation within this volume at the point where they become relevant to human reproduction. In the age of climate crises and viral pandemics, anthropocentric frameworks are increasingly inadequate on their own to either diagnose or respond to the more than non- and more than-human forces and processes that shape futures in and beyond the urban (Meehan and Strauss 2015; McKiethen and Naslund 2017). The current context of changes in the planet's systems and the role of the urban in those processes has necessitated rethinking of the relationship between social and ecological processes (Derickson 2018b, p. 427; see also Ruddick 2017), including through consideration of multispecies encounters and entanglements across various scales from the microbial to the planetary (Tsing 2015; Leiper 2017).

Reconsidering the limits of what is understood as the social also brings us to the question of the constitutive outside of social reproduction, of questions of undecidability, and of alternative conceptual schemas for understanding the social and the urban that could usefully be brought into conversation (see Peake *et al.* 2018). Thinking through the historical constitution of social reproduction, its constitutive outsides, and opening up the social may be instrumental in providing insights into how we can think through the possibilities for transformative political action in the midst of crises in social reproduction. While the following chapters, for example, address the potential of social reproduction to create material conditions of life that escape capture by capital, more needs to be done to excavate the revolutionary potential of social reproduction work through commoning practices (cf. Linebaugh 2008) to create what Caffentzis and Federici (2014) refer to as 'anti-capitalist commons' and De Angelis and Harvie (2014) as 'commoning-beyond-capital'.

Coda: Social Reproduction and the Urban During a Pandemic

As we submit this volume for publication, we have been living in what has been routinely referred to as the unprecedented time of the COVID-19 global pandemic. Each in our own distinct and interlinked ways, the authors of this chapter and editors of this book have confronted the individualizing paradoxes and isolating demands of the present moment from the vantage point of our own homes and eerily empty city streets here in Toronto. While it is important to be reflexive about how we ourselves have coped, as editors and authors of a book focused on feminist urban theory and social reproduction, we are also compelled to question the oft-mentioned phrase that we are living in unprecedented times. We ask: What exactly is unprecedented about this time? Is it unprecedented that inequality will increase? That millions will fall into poverty? That migration to cities will increase in the face of poverty? That once open cities will move to closure? That people are not able to safely access the healthcare they need because of enduring spatializations of racism? That those who suffer from ill health rooted in socio-environmental injustice will suffer in greater numbers from a novel virus? That people who are told to stay at home are not able to do so because they have no home or because their partner or parent is violent? That people will be made sick doing an underpaid and insecure job because their employer refuses to provide for basic health and safety considerations? Or that national governments and institutions alike are exploiting a crisis to institute militarized regimes of population control, to cut off access to information, to consolidate power? We could easily ask many more questions, those which address the issues that the pandemic does not so much create these calamitous conditions, but rather exposes them.

The deep systemic injustices, inequalities, and violences that have been accelerated by responses to this crisis are not new phenomenon, especially for the huge swathes of the world's population living in states governed by conservative and neo-fascist leaders, but they are surfacing with a new intensity, shining light on capitalism as the history of the separation between capital and life. The spatial organization of our lives is marked by the pain and anxiety of this separation of life and capital. By the time you read these words, 'the situation' will have again shifted enormously; the constancy of change is now more apparent than ever, as is Dr. King's call to attend to *the fierce urgency of now*. Thus, at the same time as the deeply stretched relations of social reproduction that form the warp and weft of urban everyday life are in the spotlight, we need to confront the violent re-instantiation of the 'health' of 'the economy' at the expense of everything and everyone else.

And yet the time of COVID-19 has also shown the city to be a site of ethical and political possibilities. The politics of care and connectivity that have surfaced in accounts of everyday life in cities across the globe reveal a bottom-up collective vision for helping those who lives are marginalized – refugees, immigrants, the homeless, the underpaid, targets of violence – in ways that are sustainable and speak to equality. Time will tell if there will also be a renewed politics of solidarity that arises out of these experiences. Rather than economizing, financializing, and dehumanizing society, we call for socializing and humanizing the economy, as the path by which we can reconsider, reclaim, and reconstruct our ways of being together to envision meaningful lives. This necessary re-orientation to life beyond capitalism will require reconsideration of social reproduction for years to come.

Acknowledgements

We would like to thank the contributors to this book, the anonymous reviewers, and Leeann Bennett and Mel Mikhail for their help with the bibliography and all things technical.

Notes

1 In the long history of urban scholarship, genealogies of feminist interven-tions into the urban and social reproduction can be traced back 150 years to the 1870s. Social reproduction has been the (waxing and waning) central thread of feminist urban work since the early 1970s when it was ignited by the path-breaking debates between feminist political economists. The early work of Boserup (1970) in this period, which related to Southern cities, based on a classification of different types of cities according to the presences and absences of men and women, fell between the cracks. While development feminists took up Boserup's work in relation to women's various modes of integration into development, urban feminists remained largely unaware of it, their focus being on Northern cities and the above-mentioned debates. Northern-based scholars began to amalgamate empirical studies of the gen-dered division of labour within households with feminist Marxist political economy accounts of urbanization to address the role of social reproduction in capitalism. Building on this work, urban feminists initiated a field of study of the sites and processes of social reproduction in urban place-making and

urbanization, and of the ways in which changes in spatialities and processes of social reproduction and production affect and transform the urban. The first review of this work by urban feminists came as early as 1974 (Hapgood and Getzels 1974), followed shortly by others (Hayden and Wright 1976; Wekerle 1980) (see Peake 2020 for further elaboration).

2   Although they have not stopped in their efforts to problematize and transform this intellectual erasure, feminist scholars' patience with the tenacity of this lack of engaging with questions of social reproduction has been wearing thin over several decades and is resulting, amongst other responses, with a refusal to engage with masculinist urban theory (see Katz 2006; Derickson 2018).

3   Katz's most recent definition of social reproduction falls squarely in the political economy tradition, as 'the daily and long-term reproduction of the means of production, the labor power to make them work, and the social relations that hold them in place' (Norton and Katz, 2017, p. 1).

4   The following are some of those whose contributions defined this field for a whole generation of scholars: Pat Armstrong, Hugh Armstrong, Veronica Beechey, Patricia Connelly, Maria Rosa Dalla Costa, Diane Elson, Silvia Federici, Bonnie Fox, Selma James, Martha Gimenez, Meg Luxton, Martha MacDonald, Maureen Mackintosh, Angela Miles, Maxine Molyneux, Ruth Pearson, Wally Seccombe, Lise Vogel, and Annie Whitehead.

5   Although feminist scholars did not introduce the term 'social reproduction' (it was first introduced in the 18th century), socialist feminists were responsible for developing a fully-fledged account of it (for a genealogy of the term, see Caffentzis 2002).

6   Numerous studies have shown that although some men are engaging more in domestic work, this is uneven and far from reaching equality of participation (Altintas and Sullivan 2016; Office for National Statistics 2016; Bourantani 2017; Moyser and Burlock 2018; Woodman and Cook 2019).

7   We include fields essential for social reproduction that cross the waged/unwaged work divide, such as those of childcare, domestic work, education, and healthcare (see also Pearson and Elson 2015).

8   We agree with other critical urban scholars who argue that the lack of any global agreement on a definition of the urban, the uneven pace and form of urbanization, and the incompatability of national data sets raises serious questions about the nature of the 'global' urban (see Brenner and Schmid 2014).

9   Space prevents us from even a brief overview of this literature, but see, for example, Castells (1983) on the city as a spatial unit of collective consumption, and feminist critiques of why the provision of goods and services by the state fall short of a comprehensive understanding of social reproduction.

10   The global geoeconomic transformations triggered by the financial crash have also facilitated the global rise of the right – with its associated ideologies of fascism, nationalism, populism, xenophobia, and militarism. The associated reassertion of patriarchy and misogyny, in fixing the unstable subject of woman, is also accelerating the trend to increase the burdens on women to carry the costs of social reproduction.

11   See, for example, Piketty 2015; UN-Habitat 2016; Vidal, Tjaden, and Laczko 2018.
12   Scholars have documented the feminization of migration through the transnational migration of women for care and domestic labour and the resultant creation of global care chains (Huang *et al.* 2012; Parrenas 2012; Yeates 2012). Global care chains are central to contemporary processes of social reproduction both in responding to crises in social reproduction in contexts of increasing education levels and professional employment amongst women and the withdrawal of state support for activities of social reproduction, as well as in enabling migrant women from low income countries to support the social reproduction of their families (Yeoh and Huang 2010).
13   Contemporary migration also demonstrates the increasing entanglement of paid and unpaid reproductive labour (Pearson and Elson 2015, p. 10) as a key feature of social reproduction under the conditions of neoliberal capitalism and the financialization of labour (Federici 2018; Martin 2002).

References

Ali, F. and Nagar, R. (2003). Collaboration across borders: Moving beyond positionality. *Singapore Journal of Tropical Geography* 24 (3): 356–372.

Altintas, E. and Sullivan, O. (2016). Fifty years of change updated: Cross-national gender convergence in housework. *Demographic Research* 35 (16): 455–470.

Andrucki, M., Henry, C., McKeithen, W. *et al.* (2018). Beyond binaries and boundaries in 'social reproduction'. *Environment and Planning D: Space and Society.* www.societyandspace.org/forums/beyond-binaries-and-boundaries-in-social-reproduction (accessed 1 April 2020).

Bakker, I. (2007). Social reproduction and the constitution of a gendered political economy. *New Political Economy* 12 (4): 541–556.

Beechey, V. (1977). Some notes on female wage labour in capitalist production. *Capital and Class* 1 (3): 45–66.

Benston, M. (1969). The political economy of women's liberation. *Monthly Review* 21 (4): 31–43.

Berlant, L. (2011). *Cruel Optimism*. Durham, NC: Duke University Press.

Bhattacharya, T. (2017). Introduction: Mapping social reproduction theory. In: *Social Reproduction Theory: Remapping Class, Recentering Oppression* (ed. T. Bhattacharya), 1–20. London: Pluto Press.

Boserup, E. (1970). *Woman's Role in Economic Development*. London: George Allen & Unwin.

Bourantani, E. (2017). Queering social reproduction: UK male primary carers reconfiguring care and work. *Society and Space Open Site* (31 October). https://www.societyandspace.org/articles/queering-social-reproduction-uk-male-primary-carers-reconfiguring-care-and-work (accessed 8 April 2020).

Brenner, N. (2014). *Implosions/Explosions: Towards a Study of Planetary Urbanization*. Berlin: Jovis.

Brenner, N. and Schmid, C. (2014). The 'urban age' in question. *International Journal of Urban and Regional Research* 38 (3): 731–755.

Briggs, L. (2017). *How All Politics Became Reproductive Politics: From Welfare Reform to Foreclosure to Trump*. Oakland, CA: University of California Press.

Bryan, D. and Rafferty, M. (2014). Financial derivatives as social policy beyond crisis. *Sociology* 48 (5): 887–903.

Buckley, M. and Strauss, K. (2016). With, against and beyond Lefebvre: Planetary urbanization and epistemic plurality. *Environment and Planning D: Society and Space* 34 (4): 617–636.

Burnett, P. (1973). Social change, the status of women and models of city form and development. *Antipode* 5 (3): 57–62.

Butler, J. (1993). *Bodies that Matter: On the Discursive Limits of 'Sex'*. London: Routldege.

Butler, J. (2012). Precarious life, vulnerability, and the ethics of cohabitation. *The Journal of Speculative Philosophy* 26 (2): 134–151.

Byrd, J.A., Goldstein, A., Melamed, J. *et al.* (2018). Predatory value: Economies of dispossession and disturbed relationalities. *Social Text* 36 (2 [135]): 1–18.

Caffentzis, G. (2002). On the notion of a crisis of social reproduction: A theoretical review. *The Commoner* 5. www.thecommoner.org/back-issues/issue-05-autumn-2002 (accessed 1 April 2020).

Caffentzis, G. and Federici, S. (2014). Commons against and beyond capitalism. *Community Development Journal* 49 (supplement 1): i92–i105.

Castells, M. (1983). *The City and the Grassroots: A Cross-cultural Theory of Urban Social Movements*. London: E. Arnold.

Chattopadhyay, S. (2018). Violence on bodies: Space, social reproduction, and intersectionality. *Gender, Place & Culture* 25 (9): 1295–1304.

Cixous, H. and Clément, C. (1986 [1975]). *The Newly Born Woman* (trans. B. Wing). Minneapolis, MN: University of Minnesota Press.

Daigle, M. (2019). The spectacle of reconciliation: On (the) unsettling responsibilities to Indigenous peoples in the academy. *Environment and Planning D: Society and Space* 37 (4): 703–721. https://journals.sagepub.com/doi/10.1177/0263775818824342

Daigle, M. and Ramírez, M.M. (2019). Decolonial geographies. In: *Keywords in Radical Geography: Antipode at 50* (eds. T. Jazeel, A. Kent, K. McKittrick et al.), 78–84. Hoboken, NJ: John Wiley & Sons, Ltd.

De Angelis, M. and Harvie, D. (2014). The commons. In: *The Routledge Companion to Alternative Organizations* (eds. M. Parker, G. Cheney, V. Fournier, and C. Land), 280–294. Abingdon: Routledge.

de Leeuw, S. and Hunt, S. (2018). Unsettling decolonizing geographies. *Geography Compass* 12 (7): 1–14.

Derickson, K. (2018a). Masters of the universe. *Environment and Planning D: Society and Space* 36 (3): 556–562.

Derickson, K. (2018b). Urban geography III: Anthropocene urbanism. *Progress in Human Geography* 42 (3): 425–435.

Derrida, J. (1978). *Writing and Difference* (trans. A. Bass). Chicago, IL: University of Chicago Press.

Duarte, M.E. and Belarde-Lewis, M. (2015). Imagining: Creating spaces for Indigenous ontologies. *Cataloging & Classification Quarterly* 53 (5–6): 677–702.

England, K. (1994). Getting personal: Reflexivity, positionality and feminist research. *The Professional Geographer* 46: 80–89.

Esson, J., Noxolo, P., Baxter, R. *et al.* (2017). The 2017 RGS-IBG chair's theme: Decolonising geographical knowledges, or reproducing coloniality? *Area* 49 (3): 384–388.

Fanon, F. (1961). *The Wretched of the Earth.* New York: Grove Press.

Faria, C. and Mollett, S. (2014). Critical feminist reflexivity and the politics of whiteness in the 'field'. *Gender, Place & Culture* 23 (1): 79–93. https://www.tandfonline.com/doi/abs/10.1080/0966369X.2014.958065

Faria, C. and Mollett, S. (2018). The spatialities of intersectional thinking: Fashioning feminist geographic futures. *Gender, Place & Culture* 25 (4): 565–577.

Federici, S. (2018). On reproduction as an interpretative framework for social/gender relations. *Gender, Place & Culture* 25 (9): 1391–1396.

Federici, S. (2019). On Margaret Benston: The political economy of women's liberation. *Monthly Review* 71 (4): 45. www.monthlyreview.org/2019/09/01/on-margaret-benston (accessed 8 April 2020).

Feldman, S., Geisler, C.C., and Menon, G.A. (2011). *Accumulating Insecurity: Violence and Dispossession in the Making of Everyday Life.* Athens, GA: University of Georgia Press.

Ferguson, S., Lebaron, G., Dimitrakaki, A. *et al.* (2016). Introduction. *Historical Materialism* 24 (2): 25–37.

Ferguson, S. (2020). *Women and Work: Feminism, Labour, and Social Reproduction.* London: Pluto Press.

Fernandez, B. (2018). Dispossession and the depletion of social reproduction. *Antipode* 50 (1): 142–163.

Floyd, K. (2016). Automatic subjects: Gendered labour and abstract life. *Historical Materialism* 24 (2): 61–86.

Goodfellow, T. and Owen, O. (2018). Taxation, property rights and the social contract in Lagos. *IDEAS Working Paper Series from RePEc.* www.search.proquest.com/docview/2059058830 (accessed 6 April 2020).

Hapgood, K. and Getzels, J. (1974). *Planning, Women, and Change.* Chicago, IL: American Society of Planning Officials.

Haraway, D. (1988). Situated knowledges: The science question in feminism and the privilege of partial perspective. *Feminist Studies* 14 (3): 575–599.

Harding, S. (1986). The instability of the analytical categories of feminist theory. *Signs: Journal of Women in Culture and Society* 11 (4): 645–664.

Hart, G. (2018). Relational comparison revisited: Marxist postcolonial geographies in practice. *Progress in Human Geography* 42 (3): 371–394.

Harvey, D. (2014). Cities or urbanization? In: *Implosions/Explosions:Towards a Study of Planetary Urbanization* (ed. N. Brenner), 52–66. Berlin: Jovis.

Hayden, D. (1983). Capitalism, socialism and the built environment. In: *Socialist Visions* (ed. S.R. Shalom), 59–81. Boston, MA: South End Press.

Hayden, D. and Wright, G. (1976). Architecture and urban planning. *Signs: Journal of Women in Culture and Society* 1 (4): 923–933.

Hayford, A.M. (1974). The geography of women: An historical introduction. *Antipode* 6 (2): 1–19.

Hoffman, L.M. (2014). The urban, politics and subject formation. *International Journal of Urban and Regional Research* 38 (5): 1576–1588.

Hong, G.K. (2018). Speculative surplus: Asian-American racialization and the neoliberal shift. *Social Text* 135, 36 (2): 107–122.

Hopkins, C.T. (2015). Introduction: Feminist geographies of social reproduction and race. *Women's Studies International Forum* 48: 135–140.

Huang, S., Yeoh, B.S.A., and Toyota, M. (2012). Caring for the elderly: The embodied labour of migrant care workers in Singapore. *Global Networks: A Journal of Transnational Affairs* 12 (2): 195–215.

Hunt, S. (2014). Ontologies of Indigeneity: The politics of embodying a concept. *Cultural Geographies* 21 (1): 27–32.

Jazeel, T. (2018). Urban theory with an outside. *Environment and Planning D: Society and Space* 36 (3): 405–419.

Jazeel, T. (2019). Singularity: A manifesto for incomparable geographies. *Singapore Journal of Tropical Geography* 40 (1): 5–21.

Kanes Weisman, L.K. (2000). Women's environmental rights: A manifesto. In: *Gender Space Architecture: An Interdisciplinary Introduction* (eds. J. Rendell, B. Penner and I. Borden), 6–12. London: Routledge.

Katz, C. (2001). Vagabond capitalism and the necessity of social reproduction. *Antipode* 33 (4): 709–728.

Katz, C. (2006). Messing with 'the project'. In: *David Harvey: A Critical Reader* (eds. N. Castree and D. Gregory), 234–246. Malden, MA: Blackwell Publishing.

Kim, J. (2018). Settler modernity, debt imperialism, and the necropolitics of the promise. *Social Text* 36 (2 [135]): 41–61.

Klein, N. (2007). *Shock Doctrine: The Rise of Disaster Capitalism*. Toronto: Knopf Books.

Kobayashi, A. (1994). Coloring the field: Gender, 'race', and the politics of fieldwork. *The Professional Geographer* 46 (1): 73–80.

Kofman, E. (2017). Reproduction: Social. In: *International Encyclopedia of Geography: People, the Earth, Environment and Technology* (D. Richardson, N. Castree, M.F. Goodchild et al.), pp.1–34. Oxford: John Wiley and Sons, Ltd. doi:10.1002/9781118786352.wbieg0308

Kollontai, A. (1977 [1909]). The social basis of the woman question. In: *Selected Writings of Alexandra Kollontai* (ed. and trans. A. Holt), pp. 58–73. London: Allison & Busby.

Laslett, B. and Brenner, J. (1989). Gender and social reproduction: Historical perspectives. *Annual Review of Sociology* 15: 383–384.

Lawson, V. (1995). The politics of difference: Examining the quantitative/qualitative dualism in post-structuralist feminist research. *The Professional Geographer* 47 (4): 449–457.

Lefebvre, H. (2003 [1970]). *The Urban Revolution* (trans. R. Bononno). Minneapolis, MN: University of Minnesosta Press.

Leiper, C. (2017). 'Re-wilding' the body in the Anthropocene and out ecological lives' work. *Environment and Planning D: Society and Space* (14 November). www.societyandspace.org/articles/re-wilding-the-body-in-the-anthropocene-and-our-ecological-lives-work (accessed 8 April 2020).

Linebaugh, P. (2008). *The Magna Carta Manifesto: Liberties and Commons for All*. Berkeley, CA: University of California Press.

Lofland, L. (1975). The 'thereness'of women: A selective review of urban sociology. In: *Another Voice: Feminist Perspectives on Social Life and Social Science* (eds. M. Millman and R.M. Kanter), 144–170. New York: Anchor Press.

Mackenzie, S. (1980). *Women and the Reproduction of Labour Power in the Industrial City*. Falmer, Brighton: University of Sussex (Urban and Regional Studies). http://www.worldcat.org/oclc/609593273

Maldonado-Torres, N. (2007). On the coloniality of being: Contributions to the development of a concept. *Cultural Studies* 21 (2–3): 240–270.

Markusen, A. (1980). City spatial structure, women's household work and national urban policy. *Signs: Journal of Women in Culture and Society* 5 (3): S22–S44.

Martin, R. (2002). *The Financialization of Daily Life*. Philadelphia, PA: Temple University Press.

Marx, K. (1993 [1885]). *Capital: Volume II* (trans. D. Fernbach). Middlesex: Penguin Classics Ltd.

Mbembe, A. (2004). Aesthetics of superfluity. *Public Culture* 16 (3): 373–405.

McFarlane, C. (2010). The comparative city: Knowledge, learning, urbanism. *International Journal of Urban and Regional Research* 34 (4): 725–742.

McKeithen, W. and Naslund, S. (2017). Worms and workers: Placing the more-than-human and the biological in social reproduction. *Environment and Planning D: Society and Space* (14 November). www.societyandspace.org/articles/worms-and-workers-placing-the-more-than-human-and-the-biological-in-social-reproduction (accessed 8 April 2020).

Meehan, K. and Strauss, K. eds. (2015). *Precarious Worlds: Contested Geographies of Social Reproduction*. Athens, GA: The University of Georgia Press.

Miraftab, F. (2016). *Global Heartland: Displaced Labour, Transnational Lives, and Local Placemaking*. Bloomington, IN: Indiana University Press.

Mitchell, K., Marston, S.A., and Katz, C. eds. (2004). *Life's Work: Geographies of Social Reproduction*. Malden, MA: Blackwell Publishing.

Mohanty, C. (1988). Under Western eyes: Feminist scholarship and colonial discourses. *Feminist Review* 30 (1): 61–88.

Moyser, M. and Burlock, A. (2018). Time use: Total work burden, unpaid work, and leisure. Women in Canada: A gender-based statistical report.

Statistics Canada. www150.statcan.gc.ca/n1/en/pub/89-503-x/2015001/article/54931-eng.pdf?st=30YnK5X5 (accessed 9 April 2020).

Nagar, R. (2019). *Hungry Translations: Relearning the World Through Radical Vulnerbility.* Champaign, IL: University of Illinois Press.

Nagar, R. (2002). Footloose researchers, 'traveling' theories, and the politics of transnational feminist praxis. *Gender, Place & Culture* 9 (2): 179–186.

Nagar, R. (2008). Languages of collaboration. In: *Feminisms in Geography: Rethinking Space, Place, and Knowledges* (eds. P. Moss and K.F. Al-Hindi), 120–129. Plymouth: Rowman & Littlefield.

Nijman, J. (2007). Introduction: Comparative urbanism. *Urban Geography* 28 (1): 1–6.

Norton, J. and Katz, C. (2017). Social reproduction. In: *The International Encyclopedia of Geography* (eds. D. Richardson, N. Castree, M.F. Goodchild et al.), 1–11. Oxford: Wiley & Sons, Ltd.

Noxolo, P. (2017). Introduction: Decolonising geographical knowledge. *Area* 49 (3): 317–319.

Office for National Statistics (2016). Women shoulder the responsibility of 'unpaid work'. *Office for National Statistics.* www.ons.gov.uk/employmentandlabourmarket/peopleinwork/earningsandworkinghours/articles/womenshouldertheresponsibilityofunpaidwork/2016-11-10 (accessed 9 April 2020).

Parreñas, R.S. (2012). The reproductive labour of migrant workers. *Global Networks* 12 (2): 269–275.

Peake, L. (2015). The Suzanne Mackenzie Memorial Lecture: Rethinking the politics of feminist knowledge production in Anglo-American geography. *The Canadian Geographer/Le Géographe Canadien* 59 (3): 257–266.

Peake, L. (2016). On feminism and feminist allies in urban geography. *Urban Geography* 37 (6): 830–838.

Peake, L. (2017). Feminism and the urban. In: *A Research Agenda for Cities: Elgar Research Agendas* (ed. J. Short), 82–97. Cheltenham: Edward Elgar.

Peake, L. (2020). Gender and the city. In: *International Encyclopaedia of Human Geography* (ed. A. Kobayashi), Second Edition, Vol. 5, 281–292. London: Elsevier.

Peake, L., Patrick, M., Reddy, R. *et al.* (2018). Placing planetary urbanization in other fields of vision. *Environment and Planning D: Society and Space* 36 (3): 374–386.

Pearson, R. and Elson, D. (2015). Transcending the impact of the financial crisis in the United Kingdom: Towards plan F – a feminist economic strategy. *Feminist Review* 109 (1): 8–30.

Piketty, T. (2015). Capital, inequality and justice: Reflections on capital in the twenty-first century. *Basic Income Studies* 10 (1): 141–156.

Pratt, G. (2018). One hand clapping: Notes towards a methodology for debating planetary urbanization. *Environment and Planning D: Society and Space* 36 (3): 563–569.

Puar, J.K. ed. (2012). Precarity talk: A virtual roundtable with Lauren Berlant, Judith Butler, Bojana Cvejić, Isabell Lorey, Jasbir Puar, and Ana

Vujanović. *TDR: The Drama Review* 56 (4): 163–177. https://muse.jhu.edu/article/491900

Rendell, J., Penner, B., and Borden, I. eds. (2000). *Gender Space Architecture: An Interdisciplinary Introduction.* London: Routledge.

Robinson, J. (2011). Cities in a world of cities: The comparative gesture. *International Journal of Urban and Regional Research* 35 (1): 1–23.

Roy, A. (2009). The 21st-century metropolis: New geographies of theory. *Regional Studies: The Futures of the City Region* 43 (6): 819–830.

Roy, A. (2016a). Who's afraid of postcolonial theory? *International Journal of Urban and Regional Research* 40 (1): 200–209.

Roy, A. (2016b). What is urban about critical urban theory? *Urban Geography* 37 (6): 810–823.

Ruddick, S. (2015). Situating the Anthropocene: Planetary urbanization and the anthropological machine. *Urban Geography* 36 (8): 1113–1130.

Ruddick, S. (2017). Rethinking the subject, reimagining worlds. *Dialogues in Human Geography* 7 (2): 119–139.

Ruddick, S., Peake, L., Tanyildiz, G.S. *et al.* (2018). Planetary urbanization: An urban theory for our time? *Environment and Planning D: Society and Space* 36 (3): 387–404.

Schindler, S. (2017). Towards a paradigm of Southern urbanism. *Cities* 21 (1): 47–64.

Simone, A.M. (2004). People as infrastructure: Intersecting fragments in Johannesburg. *Public Culture* 16 (3): 407–429.

Simone, A.M. (2009). *City Life from Jakarta to Dakar: Movements at the Crossroads.* London: Routledge.

Spain, D. (2002). What happened to gender relations on the way from Chicago to Los Angeles? *City and Community* 2 (2): 155–167.

Tadiar, N.X. (2013). Life-times of disposability within global neoliberalism. *Social Text* 115, 31 (2): 19–48.

Tanyildiz, G.S. (2021). *A Communism Made to the Measure of the World: A Phenomenology of the Diremption of Life and Capitalism.* Doctoral Dissertation. York University.

Tsing, A. (2015). *The Mushroom at the End of the World.* Princeton, NJ and Oxford: Princeton University Press.

Tuck, E. and Yang, K.W. (2012). Decolonization is not a metaphor. *Decolonization: Indigeneity, Education & Society* 1 (1): 1–40.

UN-Habitat (2016). *Urbanization and Development: Emerging Futures.* Nairobi: United Nations Human Settlements Programme.

Vidal, E.M., Tjaden, J.D., and Laczko, F. (2018). *Global Migration Indicators: Insights from the Global Migration Data Portal.* Berlin: Global Migration Data Analysis Centre.

Wade, R. (2009). Steering out of the crisis. *Economic and Political Weekly* 44 (13): 39–46.

Ward, K. (2010). Towards a relational comparative approach to the study of cities. *Progress in Human Geography* 34 (4): 471–487.

Ware, V. (1992). Moments of danger: Race, gender, and memories of empire. *History and Theory* 31 (4): 116–137.

Wekerle, G.R. (1980). Review essay: Women in the urban environment. *Signs: Journal of Women in Culture and Society* 5 (3): S188–S214.

Winders, J. and Smith, B.E. (2019). Social reproduction and capitalist production: A genealogy of dominant imaginaries. *Progress in Human Geography* 43 (5): 871–889.

Woodman, D. and Cook, J. (2019). The new gendered labour of synchronisation: Temporal labour in the new world of work. *Journal of Sociology* 55 (4): 762–777.

Yeates, N. (2012). Global care chains: A state-of-the-art review and future directions in care transnationalization research. *Global Networks* 12 (2): 135–154.

Yeoh, B. and Huang, S. (2010). Transnational domestic workers and the negotiation of mobility and work practices in Singapore's home-spaces. *Mobilities* 5 (2): 219–236.

2

Sociability and Social Reproduction in Times of Disaster

Exploring the Role of Expressive Urban Cultural Practices in Haiti and Puerto Rico

Nathalia Santos Ocasio and Beverley Mullings
(Queen's University)

Introduction

In an article exploring the soundscapes that emerged after the 2010 earthquake in Haiti, Elizabeth McAlister (2012) draws our attention to the ways that Haitians used music to hold themselves together through the trauma of the earthquake and its aftermath. Noting how the embodied act of singing breathed energy into Haitians, both individually and collectively, McAlister argues that 'Haitian quake survivors sang to reconstitute themselves as individuals, and to reconstitute the groups – families, neighborhoods, congregations, and communities – to which they belong' (2012, p. 25). Within months of the earthquake, a new genre of music – Rabòday – emerged within the city of Port-au-Prince, that has increasingly become a voice of critique and resistance, especially amongst young people in the city's poorest neighbourhoods. Unlike the largely religious content of the songs sung in the refugee camps that were set up in the downtown area, Rabòday, with its mixture of older Rara musical elements, electronic syncopated rhythms, and politically

A Feminist Urban Theory for our Time: Rethinking Social Reproduction and the Urban, First Edition. Edited by Linda Peake, Elsa Koleth, Gökbörü Sarp Tanyildiz, Rajyashree N. Reddy & darren patrick/dp.
© 2021 John Wiley & Sons Ltd. Published 2021 by John Wiley & Sons Ltd.

charged lyrics continues to challenge the political status quo by critiquing the ongoing event of disaster fuelled by the subordination of the island to imperial powers, corruption, and continued efforts to institute free market logics into every aspect of Haitian life. In Puerto Rico, the disaster created by hurricane María in 2017 similarly revitalized ancestral Afro-Boricua Bomba and Plena musical traditions creating new forms of sociability amongst urban residents, but also new demands for change. While the re-emergence of Plena and Bomba musical traditions in urban sites began before the hurricane, their growing popularity speaks to their importance as a space of renewal and voice. That a new generation of women's Plena groups with songs critiquing Puerto Rico's colonial status, political corruption, and economic crisis, has emerged within this typically male-dominated musical genre (Reichard 2019) speaks to the insurgent possibilities that expressive cultural practices offer in times of disaster. In this chapter we seek to explore the role of musical expressive practice in urban social reproduction in the Caribbean by documenting the role that Rabòday and Plena are increasingly playing in the maintenance of daily life in Haiti and Puerto Rico respectively. By juxtaposing these cases we explore how the return to African ancestral musical expressive practices is creating new spaces of sociability that are sustaining urban life both on a daily basis and intergenerationally, in the context of rapidly eroding physical and social urban infrastructure.

With over two-thirds of its population living in cities, the Caribbean is often categorized as one of the most urbanized regions in the world where urban life with its density of exchanges, networks, and flows characterize the majority of everyday encounters experienced by its populations. While the average annual rate of growth of the region's urban population has slowed considerably – declining steadily from 2.6% in the 1970s to 0.8% since 2010, the proportion of Caribbean people whose lives are shaped by global urban processes has continued to steadily expand (World Bank 2019). Therefore, to speak of Caribbean nation states is to speak of places thoroughly shaped by urban economic, social, and political processes and forms, but also unique plantation legacies that animate a variety of systems of oppression and unique forms of struggle against them.

The Caribbean is also one of the most vulnerable regions to the destructive effects of climate change and many of its island states are disproportionately represented amongst the 25 most-vulnerable states in terms of disasters per-capita or land area (IMF, Ötker and Srinivasan 2018). Indeed, as a recent International Monetary Fund report states, the region is seven times more likely to be hit by natural disasters than larger states, and twice as vulnerable as other small states (IMF 2018). The economic and social impacts of disasters like Goudou Goudou, the name given to the 7.0 magnitude earthquake that in 2010 devasted the cities of Port-au-Prince, Leogane, Petit Goâve, and Jacmel in Haiti, killing over 230 000 people; or Matthew, the 2016 category 4 hurricane responsible for USD$2.8 million in damage and 546 deaths across Haitian cities like Jérémie, Roche à Bateau, and Port-Au-Prince; or María, the category 5 hurricane that ravaged the islands of Dominica, the US Virgin Islands, and Puerto Rico in 2017, leaving over 3000 dead (Milken Institute School of Public Health 2018); as well as the magnitude 6.4 earthquake that shook the south of Puerto Rico on 7 January 2020, initiating an earthquake swarm that lasted weeks and precipitated the collapse of an already precarious physical and social infrastructure, highlight not only the extreme vulnerability of the region to climate change induced hazards, but also the 'not so natural' roots of these events when they occur.

Across the region, over 30 years of free market-oriented policy reform has left its imprint on urban space. Forms of urban governance premised on free market logics have been the main catalysts behind the increasing prioritization of individual over collective rights, the privatization of public space, and the formulation of new ways to extract rents from urban infrastructure. All of these forms of neoliberalization have created new patterns of uneven development and urban fragmentation across the Caribbean, deepening already existing divisions between the rich and the poor, whilst threatening the capacities of the poorest to sustain life and avoid premature death. From the securitized enclaves of private gated communities and exclusive shopping malls, to the over policed and underserved neighbourhoods of the poorest, the neoliberalization of Caribbean cities has been a key source of destruction of the types of physical and social infrastructure that in earlier years had served to narrow existing post-plantation divisions based on social class, racial hierarchies, and ethnic differences. In the context of the environmental disasters that now sweep across the region with startling regularity, the prior disinvestment in social reproduction, and the 'hostile privatism' that acompanied it (Katz 2008; Klein 2018), have exposed the much deeper threat that neoliberal forms of governance pose to the 'viability of our societies as functional entities in any meaningful sense of the word', as Norman Girvan once observed (2011, p. 1).

In Haiti, the connections between the environmental hazards and socio-economic crises that have wreaked havoc across the island state for the past 30 years can be traced back to what David Graeber (2011) calls 'permanent debt peonage' – a form of political control that has dogged the island state since its independence. With Haitian public debt ballooning from USD$43 million in 1972 to USD$750 million in 1986 (Giroud 2009), due largely to excessive unbudgeted spending for personal gain by the Duvalier regime, Haiti experienced decades of reduced public investment and compromised fiscal flexibility that eroded much of its economic and political infrastructure. Combined with the devasting impact of trade liberalization on Haitian exports, the potential for a catastrophic disaster was already in the making even before the earthquake struck. In Puerto Rico, debt and colonial political and economic control also laid the foundations for the disaster that unfolded when hurricane María struck. A model for development across the Caribbean in the 1950s, Puerto Rico's ability to attract and maintain US investments came to an abrupt end when the system of subsidies made possible through favourable federal tax laws began to be phased out in the 1990s. Attempts to compensate for declining tax revenues through government bond issues further exacerbated the problem resulting in a debt burden of USD$72 billion that, like Haiti, left the island vulnerable to outside political control. After 2016, when the US government took over financial control of the island via the Puerto Rico Oversight, Management and Economic Stability Act (PROMESA) and rolled out its package of free-market austerity policies, the island's economic and political capacity to weather the hurricane was even more compromised. While the hazard events that we document in this chapter have passed, the severe damage to the fabric of everyday life continues to unfold, and has deepened in the context of the novel coronavirus pandemic, which by 9 October 2020 had claimed the lives of 230 people in Haiti and 715 in Puerto Rico (The New York Times 2020). This is because in the absence of stability and sovereign control, both islands have become exposed to new forms of primitive accumulation led by corrupt politicians, non-governmental organizations, and a new generation of market-savvy foreign entrepreneurs intent on finding new opportunities for wealth extraction from the island's non-renewable resources. For ordinary people on both island territories, daily life has become an outrage over which few have control but for which all must assume responsibility. Making sense of how ordinary people continue to make life in situations where disaster management is a daily reality, requires an understanding of the tools that populations use to build infrastructures to sustain themselves when all else fails.

Drawing on AbdouMaliq Simone's (2004) and Georg Simmel's (1949) writings on social relationships that create and sustain the urban, and framed within social reproduction theory, we aim to identify and make clear some of the interconnections that Girvan saw lacking in contemporary approaches to disaster, and ultimately the capacity of the region to reproduce itself. In particular, we seek to make connections between the disastrous effects of debt peonage, enforced austerity, and the invasion of market logics into urban Caribbean life, and the social relationships that reconstitute 'people as infrastructure', to quote AbdouMaliq Simone (2004), in order to sustain urban life. Exploring Georg Simmel's (1949) early theorization of sociability alongside Simone's conception of infrastructure, we seek to explore the prefigurative politics embedded in expressive practices aimed at sustaining urban life in the aftermath of disaster. We seek to delineate what sociability entails, the expressive cultural practices upon which it depends, and the gender divisions of risk and responsibility that the art of human association entails (Simmel 1949). We write as scholars with ancestral and long-standing ties to the Caribbean. Our past experiences of life in the region combined with the intensity of our transnational connections as members of Caribbean diaspora communities provides a rich source of knowledge that we rely on to guide our questions, our analysis, and the secondary sources of data upon which our analysis is based.

This chapter is based on a comparative case study of Haiti and Puerto Rico and aims to explore the relationship between musical expressive practices and the social reproduction of urban communities in the aftermath of disaster. It is divided into four sections. The first examines how the work of rebuilding social infrastructure is perceived by scholars and practitioners who study and respond to disasters. More specifically, we explore what Kevin Grove (2013), calls 'the hidden transcripts of resilience'[1] in order to understand the relationship between the discourses of vulnerability and resilience currently shaping disaster management and their relationship to the social reproductive work entailed in expressive cultural practice. We follow this with a focus on the relationship between forms of sociability, expressive cultural practice, and social reproduction in the Caribbean. We pay particular attention to the ways in which the region's plantation histories and (drawing on the work of Stuart Hall) its unspoken *Présence Africaine* function as key elements of the contemporary forms of urban social infrastructure that people create to sustain themselves during periods of disaster. We then turn to forms of expressive cultural practice, more specifically music, that have (re)emerged in Haitian and Puerto Rican communities ravaged by economic and environmental disaster, and explore their performative elements and function as forms of social infrastructure. Finally, we consider the prefigurative politics that social reproduction through music offers the Caribbean. We pay particular attention to ways in which institutions such as the state or local elites currently understand these expressive cultural practices, the reasons why they seek to govern/control the social reproductive labour entailed in these forms of creative expression, and ultimately the conditions of possibility for transformational change that these practices generate.

The Hidden Transcript of Resilience and Its Social Reproductive Roots

Efforts to bring questions of vulnerability into understandings of environmental risk and disaster management, have shifted the spotlight away from the provision of scientific and technological interventions to avert disaster, towards explorations of the social relationships and structures that shape the capacity of communities to respond to disasters. As a result, a growing number of disaster scholars and practitioners now pay attention to the social relations and structures that influence how communities respond to disaster events. Cast within discourses of vulnerability and resilience, policy approaches to disaster aim to uncover the socio-economic and root causes of disasters and their impact on the ability of individuals and communities to respond to them (Thomas et al. 2013). But as a number of scholars make clear (Pelling 2003; Thomas et al. 2013), gaps and absences still remain in the way that power is conceived of, even in integrated approaches. And, as Thomas et al. (2013) observe, questions of vulnerability frequently remain 'in the position of an "add-on" while mainstream thinking, practice, and funding continue to be driven by the more traditional missions of the agencies' (Thomas et al. 2013, p. 10). There remains, for example, insufficient recognition of the economic and political risks that trade liberalization, debt, or the erosion of state investments in the social reproduction pose to the capacity of communities to recover from disaster events. Most approaches focus instead on the capacity of individuals and communities to generate their own means of recovery, even in situations where institutional goals such as poverty reduction and shared prosperity (World Bank 2013) appear as impossible goals. Indeed, it is still popular for disaster approaches to utilize language that, as McKittrick and Woods observed in the case of hurricane Katrina, serves 'to naturalize poor and black agony, distress and death' (2007, p. 7). As a number of scholars point out (Walker and Cooper 2011; Grove 2013), resilience discourses that position communities as key actors in disaster management function as a neoliberal governing technology that 'makes disaster management something each person is responsible for, rather than a matter for staff members in key state agencies' (Grove 2013, p. 199). Indeed, as Jeremy Walker and Melinda Cooper further observe, discourses of resilience are successful in downloading risk to the poorest because they are 'abstract and malleable enough to encompass the worlds of high finance, defence and urban infrastructure within a single analytic' (2011, p. 144) whilst simultaneously obfuscating their roles and responsibilities in production of disaster itself.

The shift in responsibility for social problems from the state to the individual has long been argued to be a feature of neoliberal reason (Chandler and Reid 2016). Inverting classical liberal theories that assume freedom to be a prerequisite for autonomy, neoliberal discourses define freedom as the outcome of the choices that individuals make – effectively making vulnerable subjects, instead of the structures and institutions that govern them, the cause of social problems. Discourses of resilience in disaster management seek to produce neoliberal subjects with the adaptive capacities to manage risk and uncertainty in a world where threats and dangers are beyond their control. But as Julian Reid observes, 'the technologies involved in the creation of resilient subjects are also depoliticizing ones as they require: the deliberate disabling of the political habits, tendencies and capacities of peoples and replacing them with adaptive ones' (Chandler and Reid 2016, p. 53). Thus, Reid argues that ultimately, resistance to the neoliberal governmentality bound up in discourses of resilience requires bringing into being a different kind of subject, one who is capable of imagining themselves otherwise and taking action to embrace alternative futures.

Popular critiques of vulnerability and resilience discourses (Grove 2013; Chandler and Reid 2016) generally focus on the type of subjectivities that they seek to cultivate, i.e. communities that can be trusted to be responsible and adaptable to the uncertainties and instabilities of both climate change and neoliberal capitalism. But few studies interrogate the types of embodied subjects that resilience policies are founded upon, and importantly, their location within marginalizing and intersecting systems of power. For example, feminist scholars have pointed out that the work entailed in sustaining social life does not fall evenly on all shoulders (Strauss and Meehan 2015; Teeple Hopkins 2015), and this is especially so in the aftermath of disaster (Neumayer and Plümper 2007). As Cindi Katz (2008) argues in her examination of the post-Katrina disaster recovery in New Orleans, paying attention to questions of daily life and its restoration after events like hurricanes, reveal not only how and why particular bodies come to bear the burden of responsibility for repair and recovery when a disaster unfolds, but also how the reproductive labour implicated in community-lead recovery can give rise to broader transformational politics. Drawing from social reproduction theory, Katz asks us to consider 'where the means of social reproduction have come from and with what consequences. Who has been responsible for restoring daily life, for reclaiming, reconstructing, and reviving the means of production; and for returning the provisions associated with the social wage and remaking the means of its mediation?' (Katz 2008, p. 17). To answer these questions is to make explicit the invisible and entangled relationship between work required to produce life, and work required to produce the economy.

Johanna Brenner and Barbara Laslett provide a definition of social reproduction that, though dated, still captures the various forms of labour that sustain and reproduce life. They define social reproduction as:

> the activities and attitudes, behaviors and emotions, responsibilities and relationships directly involved in the maintenance of life on a daily basis, and intergenerationally. Among other things, social reproduction includes how food, clothing, and shelter are made available for immediate consumption, the ways in which the care and socialization of children are provided, the care of the infirm and elderly, and the social organization of sexuality. Social reproduction can thus be seen to include various kinds of work – mental, manual, and emotional – aimed at providing the historically and socially, as well as biologically, defined care necessary to maintain existing life and to reproduce the next generation. (1989, pp. 382–383)

These forms of labour constitute the complex network of social processes and human relations that make capitalist systems function but are rarely imbued with value and are often unpaid. As many feminist scholars point out, to the extent that socially reproductive work is devalued so too are the bodies that carry out that labour. Thus, to the extent that women, children, racialized populations, and other historically marginalized groups are disproportionately represented in social reproductive work, so too is their labour. In the context of the growing extension of free market logics into every aspect of human life (Brown 2015), social reproduction theories raise important questions about the continued ability of poor households and communities to provide the daily and generational unpaid labour required to sustain capitalist wealth accumulation (Bakker and Gill 2003; Fraser 2017). Their burden is further compounded in the context of disaster, as resilience discourses drawing on developmentalist narratives of empowerment and transformation ultimately place the burden of coping and adaptation at their feet (Rothe 2017).

Disasters negatively impact poorer women the most because they usually have fewer resources and more responsibilities than others (Neumayer and Plümper 2007; Schuller 2010). As primary caretakers, women bear the brunt of the work involved in recovery in the aftermath of disaster to ensure that children and the elderly are cared for, that levels of public health are maintained, especially in the face of disrupted water and sanitation provisions, and that lives are sustained in the face of lost economic resources. In the aftermath of disaster, the work of social reproduction is even more difficult in cities where communities are reliant on municipal governments to restore basic urban infrastructure and where official recovery programmes like the UN's cash for work, and food for work employment programmes get taken up by men rather than women[2] (Schuller 2010; United Nations Development Fund for Women [UNIFEM] 2010).

In terms of the informal labour of social reproduction and resilience in times of disaster, AbdouMaliq Simone's (2004) proposition that we consider 'people as infrastructure' offers a useful framework to understand the 'emergency infrastructures' (Ficek 2018) that emerge when official institutions and infrastructures are no longer able to facilitate the movement of resources or services. His notion of people as infrastructure speaks to the activities of people in the city and the flexible and mobile ways that their mobilization of 'complex combinations of objects, spaces, persons, and practices' become a type of infrastructure – a 'platform providing for and reproducing life in the city' (p. 408). Simone's examination of the ways that people in African cities draw on (improvised) social infrastructures to create 'specific routes to a kind of stability and regularity' (2004, p. 410) outside those designed by the state or NGO's gestures to the limits of formal projects aimed at institutionalizing resilience. For as he observes: 'Although they bring little to the table of prospective collaboration and participate in few of the mediating structures that deter or determine how individuals interact with others, this seemingly minimalist offering – bare life – is somehow redeemed' (2004, p. 428), even if with no guarantees for transformative change in the future.

In the case of Puerto Rico, Rosa Ficek similarly draws our attention to the ways that 'people linked heterogenous networks into platforms that allowed them to survive the physical absence of infrastructures destroyed by María' (2018, p. 108) in the early weeks following the hurricane. For instance, she shows how people tapped into eclectic networks – from family members to stores, municipal oasis and natural springs – to obtain water on a regular basis. While in Haiti, women turned to time-honoured collective survival practices like extended family social supports, informal savings clubs known as Mins,[3] and beauty care services to generate economic resources to support urban life (Blouët and Bulit 2012). In these ways, the infrastructures improvised by 'people's activities in the city' ensure the social reproduction of dispossessed individuals and communities, and more critically so, during times of disaster. In the next section we explore how sociability has functioned as an emergency infrastructure, sustaining, reproducing, and replenishing life in the Caribbean, both historically, and recently in the context of disaster.

Sociability, Expressive Cultural Practice, and Social Reproduction in the Caribbean

A key characteristic of expressive cultural practices like music, dance, and art are the forms of sociability that they engender and the spaces they create for interaction and encounter across difference. Argued by Georg Simmel (1949) to be foundational to the making of society, he believed sociability in its purest form, to be an affective impulse – an interaction that individuals engaged in for their own sake rather than for a purposeful outcome. Simmel argued that like art or play, sociability was an affective reaction, driven in the first instance by a feeling of satisfaction in: 'the very fact that one is associated with others and that the solitariness of the individual is resolved into togetherness, a union with others' (1949, p. 255). While Simmel acknowledged that sociability in its purest form did not mirror the real world where inequalities and hierarchies were often insurmountable obstacles to social interaction, he nevertheless believed it was a goal that required participants to eschew established rules of conduct, to regard each other as equals and thus derive pleasure from social interaction for interaction's sake. As he claimed: 'sociability has no objective purpose, no content, no extrinsic results, it entirely depends on the personalities among whom it occurs. Its aim is nothing but the success of the sociable moment and, at most, a memory of it' (Simmel 1950, p. 45). AbdouMaliq Simone's description of association life amongst the most marginalized populations in the global South, suggests that embedded in the impulse towards sociability, is much more than abstract individual pleasure derived from play (Simone 2004, 2018). As he argues, forms of reciprocity and resource sharing amongst poor urban residents are often structured by community obligations and codified moral prescriptions, aimed at preserving a sense of coherence under conditions of precarity (2001, p. 105). And even in circumstances when forms of association are driven by divisive goals like opportunism, they are still embedded in forms of sociability that are driven not just by the pleasure of interaction, but rather, by calculations and expectations regarding individual and collective livelihoods and aspirations in the future. Implicit in Simmel's writings is a vision of urban space where strangers interact, setting aside or refusing to perform identities based on particular social relations. Sociability therefore makes it possible for collaborations across difference to take place, which, over time, have the capacity to create infrastructures that support and reproduce collectivities and communities (see also Bayón and Saraví 2013). But what Simone's work suggests, is that the impulses driving urban sociability are a complex arrangement of affective and emotional responses that are imbricated in the labour required to sustain life. For this reason, we argue that as innocuous as spaces for urban interaction through music, dance, or art may appear, each form of expressive practice holds the capacity to do more than create 'pleasurable memories'. To the extent that these practices create public spaces of encounter that sustain and socially reproduce life in the city, they should be regarded as forms of sociability that enable people to be infrastructure. Thus, just as the unpaid work involved in the provision of food, shelter, and healthcare, or in the transmission of knowledge, social values, and cultural practices (Bezanson and Luxton 2006) has come to be seen as integral to sustaining collective life, we argue, so too are the forms of sociability found in expressive labour.

Applied to the Caribbean, communities have always relied upon expressive cultural practices such as music and dance to socially reproduce themselves. Music not only exemplifies the satisfaction and pleasure that Simmel viewed as a universal impulse towards sociability, but these forms of social interaction can also be understood as a form of infrastructure, the building blocks of community that not only maintain and reproduce people, but that offer the possibility for collective and transformative change. Musical expression produces spaces for participants to create 'a sense of commonality that exceeds the sense of common loss' (Simone 2001, p. 110) that accompanies disaster, and in so doing, creates an arena for exploring new alignments and spaces for doing things differently with others.

Throughout the region's history, music and dance have served as vital sources of memory and culture that Barbara Bush (2007) notes in her exploration of the historical continuities and discontinuities in the evolution of music across the African diaspora. Constituted through shared memories of slavery and contemporary forms of racialization, Caribbean music, has also always carried a hint of subversive unpredictability. As Bush recounts: 'For Europeans, music, dance, and related cultural forms were not only a threatening reminder of the unknowable "otherness" of African slaves (and their refusal to become dehumanized chattels), but also the potential threat slaves posed to white security through obeah practices and rebellion' (2007, p. 17). Bans placed on the use of drums and horns at different periods throughout the 18th through 20th centuries in the English and Spanish Caribbean; the prohibition of dances like the Calenda, Rumba, and Bomba throughout the 20th century; and in the contemporary period, the censorship of particular Calypso ballads or Dancehall songs, all speak to the fear of subversion and transgression that these forms of expressive practice represent to ruling elites (Rath 2000; Roman 2003; Bush 2007). Rooted in African call and response traditions,[4] the ability of Caribbean music to serve as a space of cultural retention, as well as a space of creativity, resistance, and of possibility – where the unimaginable can be imagined – speaks to the social reproductive power of musical expression, especially within poor urban communities. Across the Caribbean, musical forms like Rara in Haiti, Bomba and Plena in Puerto Rico, or Reggae and Dancehall in Jamaica, offer participants a sense of connection to something larger than themselves in much the same way as Simmel describes. But we believe that participants in these forms of music also draw upon what Stuart Hall describes as the *Présence Africaine*. Drawing upon Aimé Césaire and Leopold Senghor's metaphor, Stuart Hall saw the *Présence Africaine* as the site of the repressed, invisibilized African inheritances that exist in every aspect of Caribbean life, though rarely acknowledged. As Hall notes the *Présence Africaine* , 'is "hiding" behind every verbal inflection, every narrative twist of Caribbean cultural life. It is the secret code with which every Western text was "re-read". It is the ground-bass of every rhythm and bodily movement' (1994, p. 230). It is this deeper archived set of traditions and aspirations, that communities draw upon to negotiate disasters when they threaten to damage or destroy the social infrastructure of the poorest urban communities, that we next explore.

Social Reproduction and the Unbearable Subversions of Expressive Cultural Practice: Exploring the Power of Rabòday and Plena

When the earthquake struck in 2010, Haitians were left with little more than the hidden transcripts of resilience that singing made possible, to weather the destruction that occurred. As powerful states and international development agencies gathered to decide how Haiti would be governed, rather than exposing the unfolding disaster of humanitarian governance (see Mullings, Werner, and Peake 2010), news reporters chose to focus instead on the spectacular resilience that Haitian women and children demonstrated in their production of soundscapes to sustain themselves. CNN reporters Anderson Cooper and Ivan Watson, for example, stated that:

> Despite the death and destruction, hundreds of people, mostly women, took to the streets in an area of the capital on Friday, singing and chanting as they marched down the street – a sign of resilience amid huge mounds of rubble.

It is not the first time such a display has been observed. Singing and clapping has been heard well into the night in a large square that thousands of people have made home after the earthquake. (Cooper and Watson 2010)

McAlister (2012) documents how nightly singing and the musical processions through downtown Port-Au-Prince helped people to create shared spaces of ownership in the encampments in the quake zone and, in so doing, to exercise their right to the city as a whole. Noting how: 'Repeatedly, survivors told me that their distress during the quake was so intense that they had to transform it into a song in order to withstand it' (2012, p. 25), McAlister captured at once the impulse to sociability and the social reproductive power of African musical traditions when she shared an explanation provided by one of the persons that she interviewed. Considered to be a song leader in his community, he explained that: 'If you don't have this reaction instilled in you, you cannot understand it; it's inexplicable' (2012, p. 25).

The reliance on musical expression to hold communities together has continued in the years since the earthquake and more recently after hurricane Matthew, but these newer musical forms draw heavily on older African-inspired expressive traditions that historically challenged the status quo. Rabòday has its roots in the music traditionally played during the Rara Lenten festival and sung by poor Haitians as encoded critiques of oppressive forms of power. Rara music can be traced back to the resistance movements of maroons and is often played during funeral marches, Voodoo religious services, and street protests. Looked down upon by Haitian evangelical Christians for its ties to Voodoo, and by Haitian elites because of its association with the poor, Rara music embodies the Haitian spirit of subversion in its mission to speak truth to power. Named after an already existing Rara musical style of the same name, Rabòday's popularity comes largely from its infectious rhythm but perhaps more so, for its satirical commentary and social critique. Writing for Buzzfeed News, Susanna Ferreira documents the rise of Rabòday music and its founding band – Vwadèzil and bandleader – Freshla (Ferreira 2015). She draws attention to the way that songs by the band opened up a space for public critique for discontents such as the unwelcomed presence of the UN-Peacekeeping Mission (MINUSTAH) sent by the UN Security Council to 'secure and stable environment' between 2004 and 2017 or the racial injustices perpetrated by the government of the Dominican Republic when it stripped Dominicans of Haitian ancestry of their citizenship, effectively rendering them stateless. Drawing on the tradition of satirical storytelling and the reliance on lyrics encoded with sexual double entendres, songs like 'M Pap Ka

Ba Ou Metafò'w,' that loosely translated, means either: '*I won't buy into your falsetalk or hypocrisy*' or '*I won't let you stick it in me*,' highlight the power geometries that gave rise to the unwelcomed presence of MINUSTAH on the island and that made it possible for so many women and children to be sexually abused by members of its occupying force (Ferreira 2015). As a form of expressive cultural practice, Rabòday music encourages sociability and collective critique, and is fast becoming a form of social infrastructure that is instrumental to the strategies of resistance that Haiti's poorest have begun to deploy. While some of the lyrics reproduce sexisms that reflect the everyday violence of life on the streets, much of what constitutes Rabòday creates social commentaries that disrupt and question other geometries of power.

When hurricane María made landfall in Puerto Rico, destroying critical infrastructure and killing thousands in its aftermath, Puerto Ricans also turned to expressive practices, drawing from contemporary and traditional repertoires to sustain themselves during long nights of food-sharing and safety patrols without electricity. What started in porches and parking lots (Espada-Brignoni 2018), soon moved into the broader social scene, as people gathered in bars (du Graf 2018), city squares (Univisión Noticias), and community centres (Cohen 2017) to sing, dance, and otherwise socialize and replenish themselves emotionally amidst the demoralizing scenes of destruction and the arduous work of procuring water, food, fuel, and medicine. It is in this context that Plena has seen a revival (Howard 2019). Plena is a genre associated with Puerto Rican African heritage and the working classes that originated in the early 20th century (Howard 2019). Sang to the rhythm of *panderetas*, or hand drums, Plena lyrics usually follow a basic structure of call-and-response and capture the 'current events affecting the community' (Howard 2019, p. 36). Because of its collective, interactive, and portable nature, Plena music has historically been used for street performances, Christmas celebrations, and political demonstrations. Like other cultural and expressive practices associated with the island's poorer and racialized communities, the government has always tried to control Plena by commodifying it as an old tradition devoid of restorative power, and for tourist consumption (Lloréns 2018). Before its recent boom, fans of the genre often 'were forced to attend small, late-night events in neighborhoods [mostly in the municipality of] Loíza, a coastal town with a mostly Black population, to enjoy live music' (Reichard 2019, n.p.). Despite state efforts, however, Plena has remained an important part of Puerto Rico's heritage and a pertinent cultural practice, that people turned to in order to mourn, heal, and to come together in the aftermath of hurricane María (Howard 2019; Martin 2018) and other social, political, and environmental disasters.

Like Haiti, a growing sense of abandonment and frustration amongst Puerto Ricans is likely to have contributed to the popularity of Plena Combativa, an all-women Plena group that performs old Plena songs re-interpreted from a feminist perspective as well as original songs that capture the current political and socio-economic environment in Puerto Rico. Although the group started sharing original songs through social media a few months before hurricane María made landfall, their popularity can be attributed to the attention that they bring to the multiple axes of oppression that compounded the disaster of María and its equally complex impacts on Puerto Rican lives. Their repertoire includes *Plena Indignación*, or (Plena Indignation) a song that denounces a coal-ash plant located near marginalized communities in the south of Puerto Rico, poisoning the environment and their bodies with what the lyrics call *las cenizas del capital* (the ashes of capital). Another song demands the resignation of Julia Kelleher, the now ex-secretary of Education who spearheaded the closure of over 300 public schools. As one of the verses suggests, while families were pushed out of the country after losing their community schools, the emptied buildings are now up for grabs *pa' que un riquitillo se lucre d'ella* (so a wealthy person can profit from it). Although we have shared examples only from the work of Plena Combativa here, they are one amongst many other groups and spontaneous Plenazos (Plena gatherings) that have reinvigorated the urban music scene in the years since hurricane María. By providing a space for sociability where people can let out common frustrations against the conditions that lead to the disaster of María, Plena groups are arguably helping Puerto Ricans navigate and restore daily life, just as this traditional genre did during times of struggle before. That the songs of Plena Combativa not only denounce the objectification of women (*Plena feminista para niñas felices* or *Feminist Plena for happy girls*) and the privatization of health and education (*Vivo en el país del tumbe* or *I live in the country of theft*), but also call on people to *levantarse y paralizar* (stand up and strike) speaks to the transformative potential of the social reproduction implicated in expressive practices during times of disaster.

The Possibilities and Limits of Expressive Cultural Practice to Transformational Change

In this chapter we have explored the reproductive role of sociability as mediated by expressive practices tied to African traditions in the

Caribbean in the context of disasters that threaten the sustainability of communities the region as a whole. The histories of the particular genres of expressivity that we have discussed in this chapter draw on longer African Traditions rooted in memories of oppression and resistance. Both of these musical genres respond to the 'passionate research' that Stuart Hall recovers from Fanon to discuss the cultural identities and presences of the Caribbean. Like Sonjah Stanley Niaah (2010) in her analysis of Jamaican Dancehall, we believe that these expressive practices not only offer participants tools to negotiate the poverty and violence of everyday life, but also the means to connect to a higher self and community outside of everyday routines of survival. Plena and Rabòday are providing communities with tools to hold themselves together as the free market policies that Naomi Klein (2007) describes as 'disaster capitalism' unfold.

In Puerto Rico, the collapse of the electric grid precipitated the privatization of the electric company, as well as the arrival of philantrocapitalists, cryptocurrency speculators (Klein 2018) and investors attracted by the new 'Opportunity zones' (Cintrón Arbasetti 2019) has begun to reshape the post-hurricane landscape (Lloréns 2019). As Naomi Klein's *The Battle for Paradise* (2018) reports, in March 2018, while Puerto Rican families were living without electricity or food, hundreds of cryptocurrency enthusiasts attended Puerto Crypto, a conference hosted in the luxury hotel Vanderbilt to learn about how their speculative industry could benefit from the island's corporate tax system. Although these laws pre-dated hurricane María, Klein's report showed that they were instrumental in drawing cryptocurrency investors to the island where land and public assets were being cheaply sold off, and where billions in federal disaster funds could be gained (Klein 2018, p. 21). The latter is particularly problematic in the context of the newly designated 'opportunity zones'. Opportunity zones were part of former President Trump's tax reform benefitting the wealthy, 'reduc[ing] from 37.5% to 20% the tax rate paid by funds that invest in … "low-income communities"' (Cintrón Arbasetti 2019). For Puerto Rico, the tax incentives that opportunity zones generate will likely expose 98% of the island to the vagaries of market speculators without any redistributive benefits. Furthermore, the local government added a clause exempting investors from Puerto Rico and abroad from paying construction excise taxes, while reducing municipal licence fees by 50% over a 15-year period (Cintrón Arbasetti 2019). Although the specific impacts of this new scheme are yet to be seen, the loss of revenue from construction and municipal taxes is likely to further hinder the capacity of the government to provide critical social and physical infrastructure.

While in Haiti, the post-disaster government mantra that Haiti was 'Open for business' (*The Economist* 2012) masked the continued erosion of the social infrastructures supporting everyday life, even as plans to restore the economy were boldly declared. Thus, despite the promises of international NGOs and the Interim Haiti Recovery Commission to help Haiti to 'build back better', almost a decade later it has become clear that much of the USD$9 billion pledged by international donors was either never given, inefficiently spent, or simply siphoned off by organizations and corporations intent on capitalizing on the opportunities for quick profits. A case in point is the USD$300 million USAID created Caracol Industrial Park, run by a subsidiary of Sae-A Trading Co. Ltd – a Korean clothing manufacturer known for its troubled labour relations in Guatemala. Celebrated at the time of construction for the 20 000 jobs it would create, it is now clear that the Caracol Plant not only failed to become the growth pole that it was envisioned to be, but perhaps more egregiously, its construction came at the expense of over 366 farmers who were dispossessed of their land, and over 4000 people who lost their livelihoods when the fertile tract of land upon which they farmed was appropriated to make the way for the industrial park (Accountability Counsel [AREDE] and ActionAid Haiti 2018). Action Aid described the construction of the industrial park as a case of land grabbing, because up until 2017, when a complaint was filed by a coalition of Haitian farmers with the Inter-American Development Bank, few farmers had been compensated. But the case also demonstrated the extent to which the neoliberal forms of post-disaster management adopted by the United States and Haitian Government prioritized the promise of economic gains from low-waged manufacturing over the certainty of the social reproductive gains from local food sovereignty. The connections are clearly articulated by Etienne, one of the farmers who lost his land, who states: 'It was very fertile soil. I was able to grow corn, peanuts and black beans and got two harvests a year. Now I have to buy food imported from the Dominican Republic which is much more expensive. Before corn was 4–5 gourdes per kilo; now it's 100' (ActionAid, n.d.).

While expressive practices like music and dance are a form of infrastructure that can fortify and hold communities together, they cannot do so infinitely, if the sociabilities they enable are not infused with insurgent possibilities. Indeed, we recognize that the extraordinary reproductive labour that expressive practices require might simply represent a 'negotiation of (re)newed marginality' (Black 2014, p. 702) that enables the very discourses of resilience that so many aid agencies currently applaud.

Yet, as the case of Plena Combativa makes clear, Plena music has become an instrumental part of the insurgent projects emerging in Puerto Rico. Not only has the group given communities musical tools to reflect upon and denounce the environmental harm of coal ash, school closures, and corruption, it has also explicitly invited Puerto Ricans to stand and strike against austerity through songs like *Paraliza* (*Strike*). While writing this chapter in the summer of 2019, Puerto Rico became embroiled in an historic process of mass mobilization that was successful in its demand for the resignation of governor Ricardo Rosselló Nevares (Colón Almenas et al. 2019). The mobilizations began after the FBI arrested the former heads of the Department of Education and the Health Insurance Administration (ASES) on corruption charges at the same time that close to 900 pages of a private chat between the governor and cabinet members, private consultants, and friends, all of whom were men, was leaked. In this now infamous chat, the men casually discussed public policy, made misogynistic, sexist, and homophobic comments, and joked about the high number of corpses waiting to be dispatched from the morgues after hurricane María (Minet and Valentín Ortiz 2019).

Recent mobilizations have increasingly denounced systemic corruption and questioned the validity of the debt issued by Puerto Rico.[5] In Haiti, popular demonstrations in 2019 also denounced the widespread corruption that led to the misuse of USD$3.6bn of funds earmarked for local development under Venezuela's PetroCaribe[6] oil/aid programme. Plena and Rabòday music have been notably present in each protest – creating a soundscape for people to articulate their anger and their unity, and a collective space to remember, to quote Franz Fanon: 'some very beautiful and splendid era whose existence rehabilitates ...' in ways that go beyond the misery of the present (Fanon quoted in Hall 1994, p. 223). In the case of Puerto Rico, the regular presence of Plena Combativa has become one of the many circles where people assemble to struggle. Although we cannot make definitive connections between the ongoing political mobilizations and the imaginaries that these emerging forms of musical expression ignite, we believe that the subversive lyrics of groups like Vwadèzil and Plena Combativa hint at their insurgent possibilities.

Acknowledgements

We would like to thank Linda Peake, Rajyashree Reddy, and Gökbörü Tanyildiz for their generous and critical comments on an earlier draft of this chapter and to members of the audience at the Feminist Explorations of Urban Futures Conference at York University 26–28 September 2019, where we shared some of the ideas in this chapter, for their insightful comments and questions. Any errors or oversights are, of course, ours entirely.

Notes

1   Grove's phrase 'Hidden Transcripts of Resilience' draws on the title of James C. Scott's earlier book *Weapons of the Weak: Everyday forms of peasant resistance*, in order to highlight the physical and psychic spaces of resistance against otherwise overwhelming domination that are embedded in seemingly banal everyday practices and social interactions amongst the poor.

2   UN Development Fund for Women [UNIFEM] (2010) notes that in Haiti, of the 200 000 Haitians in the first half of 2010 that benefitted from the cash-for-work and food-for work programmes only 35% were women, while Mark Schuller notes how the practice of selecting management committees to distribute aid largely excluded local women.

3   A Min is a type of Rotational Savings and Credit Association (ROSCA), commonly practised across West Africa and the Caribbean that is also known as 'Sou-Sou', 'boxhand,' or 'partner'.

4   Call and response is a form of social interaction, in which a statement made by a musician or group of musicians is answered by a response from the listeners. It is a form of sociability insofar as it invites dialogue and generates innovation and collaboration amongst participants.

5   The Financial Oversight and Management Board, locally referred to as *La Junta*, was created to oversee the restructuring of the Puerto Rican debt and manage the finances and the budget of the island. One of the most popular chants during the recent mobilizations was: 'Ricky, renuncia, y llevate a la Junta' ('Ricky, resign, and take la Junta with you') (see Aronoff 2019).

6   PetroCaribe is a strategic oil alliance between Venezuela and 17 Caribbean States that facilitates the purchase of oil at preferential rates. Under the scheme, oil-dependent Caribbean countries were able to buy fuel from Venezuela and defer payment for up to 25 years.

References

Accountability Counsel AREDE and ActionAid Haiti (2018). Haitian farmers harmed by Caracol Industrial Park reach historic agreement (10 December). www.medium.com/@AccountCounsel/haitian-farmers-harmed-by-caracol-industrial-park-reach-historic-agreement-76f4919edbff (accessed 10 July 2019).

ActionAid (n.d.). Land grabbing in Haiti: The Caracol Industrial Park. www.actionaidusa.org/work/land-grabbing-in-haiti-the-caracol-industrial-park (accessed 10 July 2019).

Aronoff, K. (2019). As Puerto Rico erupts in protests and Governor resigns, 'La Junta' eyes more Power. *The Intercept* (24 July). www.theintercept.com/2019/07/24/puerto-rico-protests-ricardo-rossello-la-junta (accessed 10 July 2019).

Bakker, I. and Gill, S. (2003). *Power, Production, and Social Reproduction: Human In/Security in the Global Political Economy*. New York: Palgrave Macmillan.

Bayón, M. and Saraví, G. (2013). The cultural dimensions of urban fragmentation: Segregation, sociability, and inequality in Mexico City. *Latin American Perspectives* 40 (2): 35–52.

Bezanson, K. and Luxton, M. (2006). Introduction: Social reproduction and feminist political economy. In: *Social Reproduction: Feminist Political Economy Challenges Neo-Liberalism* (eds. K. Bezanson and M. Luxton), 3–10. Montreal: McGill-Queen's University Press.

Black, S. (2014). Street music', urban ethnography and ghettoized communities. *International Journal of Urban and Regional Research* 38 (2): 700–705.

Blouët, A.T. and Bulit, G. (2012). Impact assessment study on the living conditions of women and children, Haïti. *Field Actions Science Reports Online* 5: 1–8.

Brown, W. (2015). *Undoing the Demos: Neoliberalism's Stealth Revolution*. New York: Zone Books.

Bush, B. (2007). African echoes, modern fusions: Caribbean music, identity and resistance in the African diaspora. *Music Reference Services Quarterly* 10 (1): 17–35. https://journals-scholarsportal-info.proxy.queensu.ca/details/10588167/v10i0001/17_aemfcmaritad.xml

Caldeira, T.P.R. (2015). Social movements, cultural production, and protests. *Current Anthropology* 56 (S11): S126–S136.

Chandler, D. and Reid, J. (2016). *The Neoliberal Subject: Resilience, Adaptation and Vulnerability*. Lanham, MD: Rowman & Littlefield.

Cintrón Arbasetti, J. (2019). El Gobierno no tiene un plan para integrar a las comunidades designadas como 'Zonas de Oportunidad'. *Centro de Periodismo Investigativo* (March 1). www.periodismoinvestigativo.com/2019/03/el-gobierno-no-tiene-un-plan-para-integrar-a-las-comunidades-designadas-como-zona-de-oportunidad (accessed 10 July 2019).

Cohen, J. (2017). 'Necesitamos La Música': Puerto Ricans recovering from María Embrace the Arts. *National Public Radio* (December 15). www.npr.org/2017/12/15/570645257/necesitamos-la-m-sica-puerto-ricans-recovering-from-maria-embrace-the-arts (accessed 10 July 2019).

Colón Alemnas, V., Minet, C., Candelas, L. et al. (2019). 934 días en la Fortaleza. *Centro de Periodismo Investigativo* (25 July). www.periodismoinvestigativo.com/2019/07/934-dias-en-la-fortaleza (accessed 10 July 2019).

Cooper, A. and Watson, I. (2010). Desperation grows: Mass grave found outside Port-au-Prince. *CNN* January 16 2010. https://www.cnn.com/2010/WORLD/americas/01/15/haiti.earthquake/index.html (accessed 1 April 2021).

Danticat, E. (2017). A new chapter for the disastrous United Nations mission in Haiti? *The New Yorker* 19 October 2017. https://www.newyorker.com/news/news-desk/a-new-chapter-for-the-disastrous-united-nations-mission-in-haiti

du Graf, L. (2018). Bomba: The enduring anthem of Puerto Rico. *The New York Times* (7 July). www.nytimes.com/2018/07/07/style/bomba-puerto-rico-music-dance.html (accessed 10 July 2019).

Espada-Brignoni, T. (2018). From the roars of hurricanes to the chords of standards: How we used popular music in the aftermath of Hurricane María in Puerto Rico. *Popular Music and Society* 42 (1): 118–122.

Ferreira, S. (2015). How disaster and tragedy spawned a radical music movement in Haiti. *BuzzFeed News* (7 July). www.buzzfeednews.com/article/susanaferreira/haiti-raboday-revolution (accessed 10 July 2019).

Ficek, R.E. (2018). Infrastructure and colonial difference in Puerto Rico after Hurricane María. *Transforming Anthropology* 26 (2): 102–117.

Fraser, N. (2017). Crisis of care? On the social-reproductive contradictions of contemporary capitalism. In: *Social Reproduction Theory: Remapping Class, Recentering Oppression* (ed. T. Bhattacharya), 21–36. London: Verso.

Giroud, S. (2009). Baby Doc's odious debts and Haiti's legal defences. In: *How to Challenge Illegitimate Debt: Theory and Legal Case Studies* (eds. M. Mader and A. Rothenbühler), 97–102. Basel: Aktion Finanzplatz Schweiz.

Girvan, N. (2011). Existential threats: Regionalising governance, democratizing politics. *CLR James Memorial Lecture*, Oilfield Workers Trade Union.

Graeber, D. (2011). *Debt: The First 5,000 Years*. New York: Melville House.

Grove, K. (2013). Hidden transcripts of resilience: Power and politics in Jamaican disaster management. *Resilience* 1 (3): 193–209.

Hall, S. (1994). Cultural identity and diaspora. In: *Identity: Community, Culture, Difference* (ed. J. Rutherford), 222–237. London: Lawrence & Wishart.

Howard, K. (2019). Puerto Rican plena: The power of a song. *General Music Today* 32 (2): 36–39.

Katz, C. (2008). The GPC Jan Monk Distinguished Lecture. Bad elements: Katrina and the scoured landscape of social reproduction. *Gender, Place & Culture* 15 (1): 15–29.

Klein, N. (2007). *Shock Doctrine: The Rise of Disaster Capitalism*. Toronto: Knopf Books.

Klein, N. (2018). *The Battle for Paradise: Puerto Rico Takes on the Disaster Capitalists*. Chicago: Haymarket Books.

Lamotte, M. (2014). Rebels without a pause: Hip-hop and resistance in the city. *International Journal of Urban and Regional Research* 38 (2): 686–694.

Laslett, B. and Brenner, J. (1989). Gender and social reproduction: Historical perspectives. *Annual Review of Sociology* 15: 381–404.

Lloréns, H. (2018). Beyond *blanqueamiento*: Black affirmation in contemporary Puerto Rico. *Latin American and Caribbean Ethnic Studies* 13 (2): 157–178.

Lloréns, H. (2019). The race of disaster: Black communities and the crisis in Puerto Rico. *Black Perspectives* (17 April). www.aaihs.org/the-race-of-disaster-black-communities-and-the-crisis-in-puerto-rico (accessed 10 July 2019).

Martin, M. (2018). Puerto Rico's devastation permeates Plena. *All Things Considered* (2 June). hwww.npr.org/2018/06/02/616472663/puerto-rico-s-devastation-permeates-plena (accessed 10 July 2019).

McAlister, E. (2012). Soundscapes of disaster and humanitarianism: Survival singing, relief telethons, and the Haiti earthquake. *Small Axe* 16 (3): 22–38.

McKittrick, K. and Woods, C. (2007). No one knows the mysteries at the bottom of the ocean. In: *Black Geographies and the Politics of Place* (eds. K. McKittrick and C. Woods), 1–13. Toronto: Between the Lines.

Milken Institute School of Public Health (2018). Ascertainment of the estimated excess mortality from Hurricane Maria in Puerto Rico. *George Washington University Milken Institute School fo Public Health*. www.publichealth.gwu.edu/content/gw-report-delivers-recommendations-aimed-preparing-puerto-rico-hurricane-season (accessed 10 July 2019).

Minet, C. and Ortiz, V. (2019). Las 889 páginas de Telegram entre Rosselló Nevares y sus allegados. *Centro de Periodismo Investigativo* (13 July). www.periodismoinvestigativo.com/2019/07/las-889-paginas-de-telegram-entre-rossello-nevares-y-sus-allegados (accessed 10 July 2019).

Mullings, B., Werner, M., and Peake, L. (2010). Fear and loathing in Haiti: Race and the politics of humanitarian dispossession. *ACME* 9 (3): 282–300.

Neumayer, E. and Plümper, T. (2007). The gendered nature of natural disasters: The impact of catastrophic events on the gender gap in life expectancy, 1981–2002. *Annals of the Association of American Geographers* 97 (3): 551–566.

'Open for Business: The new president wants to change his country's image' (2012). *The Economist* (7 January). www.economist.com/the-americas/2012/01/07/open-for-business (accessed 14 October 2020).

Ötker, İ. and Srinivasan, K. (2018). Bracing for the storm: For the Caribbean, building resilience is a matter of survival. *Finance and Development* 55(1): 49–51. www.imf.org/external/pubs/ft/fandd/2018/03/otker.htm (accessed 10 July 2019).

Pelling, M. (2003). *The Vulnerability of Cities: Social Resilience and Natural Disaster*. London: Earthscan.

Rath, R.C. (2000). Drums and power: Ways of Creolizing music in coastal South Carolina and Georgia, 1730–1790. In: *Creolization in the Americas: Cultural Adaptations to the New World* (eds. S. Reinhardt and D. Buisseret), 99–130. Arlington: Texas A&M Press.

Reichard, R. (2019) In Puerto Rico, a new generation of women's plena groups are raising their voices. *Remezcla* (18 April). www.remezcla.com/features/music/plena-combativa-women-plena-profile (accessed 10 July 2019).

Román, R.L. (2003). Scandalous Race: Garveyism, the bomba, and the discourse of blackness in 1920s Puerto Rico. *Caribbean Studies* 31 (1): 213–259.

Rothe, D. (2017). Gendering resilience: Myths and stereotypes in the discourse on climate-induced migration. *Global Policy* 8 (Suppl. 1): 40–47.

Schuller, M. (2010). Double victims: The earthquake through Haitian women's eyes. *Huffington Post* (13 April). www.huffpost.com/entry/double-victims-the-earthq_b_534974 (accessed 10 July 2019).

Simmel, G. (1949). The sociology of sociability (Trans. E.C. Hughes). *American Journal of Sociology* 55 (3): 254–261.

Simmel, G. (1950). *The Sociology of Georg Simmel*. (Trans. K.H. Wolff). New York: The Free Press.

Simone, A.M. (2001). Straddling the divides: Remaking associational life in the informal African city. *International Journal of Urban & Regional Research* 25 (1): 102–117.

Simone, A.M. (2004). People as infrastructure: Intersecting fragments in Johannesburg. *Public Culture* 16 (3): 407–429.

Simone, A.M. (2018). *Improvised Lives: Rhythms of Endurance in an Urban South*. London: Wiley.

Stanley Niaah, S. (2010). *Dancehall: From Slave Ship to Ghetto*. Ottawa: University of Ottawa Press.

Strauss, K. and Meehan, K. (2015). Introduction: New frontiers in life's work. In: *Precarious Worlds: Contested Geographies of Social Reproduction* (eds. K. Meehan and K. Strauss), 1–22. Athens GA: University of Georgia Press.

Teeple Hopkins, C. (2015). Introduction: Feminist geographies of social reproduction and race. *Women's Studies International Forum* 48: 135–140.

The New York Times (2020). Covid world map: Tracking the global outbreak. *The New York Times*. www.nytimes.com/interactive/2020/us/coronavirus-us-cases.html#states (accessed October 9, 2020).

Thomas, D.S.K., Phillips, B.D., Lovekamp, W.E. et al. (2013). *Social Vulnerability to Disasters*. Second edition. Hoboken: CRC Press.

UNIFEM (2010). At a glance: Women in Haiti. *ReliefWeb* (30 July). www.reliefweb.int/report/haiti/glance-women-haiti (accessed 30 August 2019).

Univisión Noticias (2017). *Plena Después De María: Música Puertorriqueña Para Olvidar Los Estragos Del Huracán* (24 December). www.youtube.com/watch?v=fWoMU1RgbpE (accessed 10 July 2019).

Walker, J. and Cooper, M. (2011). Genealogies of resilience: From systems ecology to the political economy of crisis adaptation. *Security Dialogue* 42 (2): 143–160.

World Bank (2013). *Building Resilience: Integrating Climate and Disaster Risk into Development*. Washington, DC, World Bank.

World Bank (2019). DataBank: World development indicators. www.data bank.worldbank.org/source/world-development-indicators# (accessed 11 September 2019).

3

'Never/Again'

Reading the Qayqayt Nation and New Westminster in Public Poetry Installations

Emily Fedoruk (University of Minnesota)

Introduction

Beginning at a point of illegibility might seem a foolish undertaking for any extended reading of poetry, and I will readily admit to being no more equipped to scan the quotation from Qayqayt Chief Rhonda Larrabee than any other passer-by who chances to see the text in its location on Columbia Street in New Westminster, British Columbia, Canada, visible from vantage points on Eighth Street such as those illustrated in Figures 3.1 and 3.2.

This 'poem' appears on unceded territories of xʷməθkʷəy̓əm (Musqueam), qiqét (Qayqayt), səlililwətaʔɬ (Tsleil-Waututh), Sḵwx̱wú7mesh (Squamish), Q'e'yc'ey (Katzie), and Qw'ʔntl'en ... (Kwantlen), part of a 2010 redevelopment project by VIA Architecture named Plaza 88 that consists of four residential towers and a major transit station.[1] The city settler Canadians called New Westminster is located 12 miles southeast from the place that came to be known as Vancouver, positioned on an historic trade route and a residence for 71 000 people (Statistics Canada 2016). I think of New Westminster as home but I am a visitor there, just as I am while finishing my dissertation in Minneapolis, Minnesota, on Dakota and Anishinabek land. My reading of Larrabee's poem considers how creative practices contribute to social reproduction in public urban space. As an Indigenous poem in the context of a settler colonial

A Feminist Urban Theory for our Time: Rethinking Social Reproduction and the Urban, First Edition. Edited by Linda Peake, Elsa Koleth, Gökbörü Sarp Tanyildiz, Rajyashree N. Reddy & darren patrick/dp.

Figure 3.1   Eastern elevation: Plaza 88, New Westminster BC. Mural quoting Chief Rhonda Larrabee (Qayqayt). (August 2017).

urban landscape, this text offers up a framework for social reproduction in that it presents something at once 'illegible' to capital and simultaneously necessary for quotidian capitalist production. But the literary ambiguity in the line feels as though it is not for the reader to discover. Larrabee's poem makes me realize that before making any statements on the possibilities for public poetry and its relations to social reproduction, I want to pause, forestalling a reading of this poem that is coherent or straightforward or somehow otherwise mine, merely because it appears in the dissertation I am writing. If I can, I want, instead, an approach to this poetic wall that respects certain boundaries of meaning, and of power, for Larrabee – and for her mother, Marie Joseph Bandura, to whom the original quotation can be attributed.

No one can actually read Larrabee's poem on this building. Across a checkerboard pattern of fragments of alphabetic letters painted on the façade of an office for an HR consulting firm, the line could say, 'I will tell you once, but you must never ask me again,' and this phrase is the text I read in this chapter. I point to possibilities for this poem to complicate concepts of social reproduction and feminism for the Qayqayt precisely because it is illegible, and furthermore, *does not* straightforwardly quote from, name, or identify Larrabee and her mother, Marie Joseph Bandura.

Figure 3.2   Eastern elevation, Plaza 88: Entrance to New Westminster Skytrain Station. (July 2018, photo by Meghan Armstrong).

Here I depend upon the politics of recognition articulated by Glen Coulthard, referring to 'the now expansive range of recognition-based models of liberal pluralism that seek to "reconcile" Indigenous assertions of nationhood with settler-state sovereignty via the accommodation of Indigenous identity claims in some form of renewed legal and political relationship with the Canadian state' (2014, p. 3). Considering my own position as a settler-reader of the poem, I read it alongside another text on the walls at Plaza 88: a line about nationalism and development written by the building's architect, Graham McGarva, that adopts a colonial voice. These texts demonstrate a crisis, I argue, of site-specificity in colonized North American urban space today. Plaza 88 cannot offer us 'historical' documents; Coulthard cites several authors, including Silvia Federici, who have pointed to 'the *persistent* role that unconcealed, violent dispossession continues to play in the reproduction of colonial and capitalist social relations in both the domestic and global contexts,' also arguing against Marx's 'rigidly *temporal* framing' of primitive accumulation (p. 9). As prime mediator of this 'violent dispossession', urban space conditions ongoing oppressions – not only displacements – as conflicts bound up in global capitalism and state recognition.

In this chapter, I read the poem into the particular urban context of the creative city, eking out three valences for social reproduction in, and around, the poem itself. Identifying the first valence through Silvia Federici's definitions of social reproduction, I position the process she describes – one by which women's unwaged labour is 'mystified' as 'a personal service and even a natural resource' – amongst Indigenous feminisms that have recently been put forward in relation to the settler Canadian nation state (2004, p. 18). The second valence of social reproduction I see in the mural is its status as artistic reproduction, but this status is equally contingent upon the contradictory third valence; the poem is itself a commodity that lends value to Plaza 88. Therefore, I read the poem as it also circumscribes limitations on urban social reproduction – limits that are thrown into relief by considering Indigenous feminisms. I attempt to articulate some of the complexities of authorship that come to the fore in this poetic text, considering my own position as a reader and parsing some of the history of the Qayqayt Nation. Chief Rhonda Larrabee, as one of the text's authors or original sources, shares this history in the settler space of New Westminster, so any act of recounting the Qayqayt story is itself a mode of social reproduction for the Nation entirely contingent upon her efforts. Finally, returning to Coulthard, I read the poem alongside the other major text 'published' by Graham McGarva at Plaza 88, a distinctly colonial narrative that documents what Coulthard calls 'the asymmetrical exchange of mediated forms of state recognition and accommodation' today and contemplate the building's situation as monument (p. 15).

Social Reproduction and the Urban in the Context of Settler Colonialism

Artist and critic Martha Rosler critiques 'the platitudinous banality of [urban studies theorist Richard] Florida's city vision, its undialectical quality and its erasure of difference in favor of tranquility and predictability as it instantiates as policy the infantile dream of perpetually creating oneself anew' (2013, p. 119). Rosler suggests that Florida's famous 'Creative City Thesis' offers for 'second-tier cities' a method to 'glorify the accumulation of amenities as a means of salvation from an undistinguished history, a chance to develop and establish flexibility' (p. 113). As a suburb of Vancouver, New Westminster could not necessarily be categorized as a second-tier city and, furthermore, its 'accumulation of amenities' does not merely describe a process of purchasing facilities to fill in the blank spaces of 'an undistinguished history' – this storeyed, self-indulgent suburb has several hallowed 'Historic Districts' and chronicles of its colonial and commonwealth lineage abound. Instead, at the recent urban redevelopment of Plaza 88, the 'glorification of accumulation' occurs in the strategic redeployment of historical information towards 'a chance to develop and establish flexibility' – as an urban determinant and economic possibility. The unevenness produced by the façades at this transit mall involve a synthetic incorporation of art practice within the fixed capital of urban development, but this site also plays into the 'platitudinous banality' Rosler experiences in Florida by committing a concrete and visible 'erasure of difference in favor of tranquility and predictability' (p. 119).

Reading this poem within the context of creativity in urban space today positions the poem as product of the economic relationship between art and the built environment. In dialogue with Rosler, I approach urban space as an historical, economic, and ideological categorization as well as within a site-specific framework for reading this poem.[2] The poem itself exists at the limits of the urban because its site-specificity is suspect; it remains an Indigenous text on stolen land. In order for Coulthard to carry out what he calls a 'contextual shift' in Marxist theory – centering the colonial relation of dispossession rather than the capitalist relation concentrated on the 'expropriation of the worker' – he seeks to 'transform' Marxism *in conversation with the critical thought and practices of Indigenous peoples themselves' (p. 8). This 'contextual shift' is specifically premised on a formulation from Silvia Federici in *Caliban and the Witch* where she writes, 'Whereas Marx examines primitive accumulation from the viewpoint of the waged male proletariat and the development of commodity production, I examine it from the changes it introduced in the social position of women and the production of labour-power' (2004, p. 12). Federici's endnote on these 'changes' for the 'social position of women and the production of labour-power' helps me to establish some of the power dynamics for Qayqayt women that are also at work in the mural:

> These two realities, in my analysis, are closely connected, since in capitalism reproducing workers on a generational basis and regenerating daily their capacity to work has become 'women's labour', though mystified, because of its un-waged condition, as a personal service and *even a natural resource.* (2004, p. 18, emphasis added)

As the first valence for social reproduction, I want to follow Federici to further decode how this mystification – as a necessary element of social reproduction – affects any opportunity to see and read Larrabee's poem as bound up in social reproduction for the Qayqayt as well as for settlers reading it in New Westminster. I understand mystification, and by extension the poem's commodification here, as inextricably linked to violence: the mural *qua* commodity captures the violent dispossession of lands from the Qayqayt and, as Sarah de Leeuw and Sarah Hunt argue, 'Racialized gender violence is integrally related to Indigenous struggles over land and self-determination, not only in the US and Canada, but globally' (2018, p. 2). We know well from Marx that a commodity is constituted by *disappearances*: the abstraction of labour is intrinsic to fetishism or in any broader mystification, to use Federici's term, thus tethering this disappearance to social reproduction. The 'two realities' that Federici joins together here of course expose something else that is lost in Marx, and that resonates in Hunt and de Leeuw: how the changing 'social position of women' and 'the production of labour power' are indistinct. 'Women's labour' is what makes possible 'reproducing workers on a generational basis' – a claim that, again, takes on more power in relation to Larrabee's Indigeneity and her mother's revelation about their family history – and effectuates the means of 'regenerating daily their capacity to work' (Federici 2004, p. 18). This recognition gets at a much deeper, lengthier, and more difficult disappearance within the so-called abstraction of labour; capitalism is contingent upon the historical and ongoing violence of heteropatriarchal colonialism, chasing labour power around the globe, and fixed to what Hunt and de Leeuw call 'racialized gender violence' (p. 2). Federici also shows social reproduction lapped up as 'even a natural resource' (p. 18). This characterization complicates the lines Coulthard draws between labour and dispossession – in another key revision, considering the primacy of land in Canadian colonialism, he suggests that 'the history and experience of *dispossession*, not proletarianization, has been the dominant background structure shaping the character of the historical relationship between Indigenous peoples and the Canadian state' (p. 9). 'As a natural resource', social reproduction is premised on still more acts of violent '*dispossession*', and lands and labour are inextricable.

But in this context of so many abstractions, *is* Larrabee's poem social reproduction? Or does it splice into commodification at some other point of mystification? The poem is, of course, an *artistic* reproduction, its second valence of social reproduction: McGarva, Plaza 88's architect, took the line from Larrabee's conversation and designed a mural – which, as I show, he identifies as a poem when discussing the project in local media – on the side of one of his new developments. Furthermore, it is immediately obvious how the line 'I will tell you once, but you must never ask me again,' as a public instance of retelling, thus exposes the terms of exploitation and expropriation that are bound up in the Qayqayt's particular struggle for social reproduction as a Nation erased in the settler Canadian socio-political context – and the urban space of New Westminster – until the 1990s. In this sense, social reproduction is at stake for the

Qayqayt Nation but is also carried out by settler-readers of the poem – including my own reading here, and tied to my own knowledge production in the academy. A third valence of social reproduction would be the contradictions involved in the poem's awkward position as a commodity. If I read and view the poem as it brands, adds value, or even decorates this mall, how can I reconcile its reproductive status? How are these public poetry installations useful to the capital, especially as, in this case, the poem cannot be read? I argue for poetry's unique capacity to offer a non-accumulative mode for engaging the urban environment. Readers cannot easily follow these inscriptions as signs or an apparatus for navigation, whereas so many poems have been authored and published to promote contemplation and self-reflection in the urban environment.[3] But even this contemplation is premised on abstraction, because the reader is expected to engage in an act of meaning-making that requires decoding symbols, using one's imagination, or, to employ a crucial term in contemporary urbanism, poems demand *creativity*.

In its publication at Plaza 88, Larrabee's quotation is at once a significant vehicle for social reproduction and, more perplexingly, a message that is being used for capitalist valuation. The poem reminds us, in both these capacities, of how capitalism perpetually exploits women's bodies. Within the North American urban context, every signal to this exploitation is a reminder of missing and murdered Indigenous women who are being disappeared across the continent, and globally – to return to Hunt and de Leeuw, 'racialized gender violence is integrally related to Indigenous struggles over land and self-determination' (p. 2). Because this poem is published in the urban context of the Lower Mainland, its representation and/or abstraction of Larrabee and her mother as Indigenous women must be considered in relation to activist work over the past several decades that has drawn public attention to missing and murdered Indigenous women in Vancouver's Downtown Eastside neighbourhood, Northern British Columbia, and too many other places in the province (Murakami 2008; Dean 2015; Eng 2016). At the same time, as poet Mercedes Eng records in her collection *Prison Industrial Complex Explodes*, there has been an 80% increase in federally sentenced Indigenous women inmates in Canada over the past decade (2017, p. 79).

Anishinaabe feminist scholar Dory Nason's 2013 essay, 'We Hold Our Hands Up: On Indigenous Women's Love and Resistance,' describes how 'women in the #IdleNoMore movement seek to protect the waters, the environment and the land from the threat of further destruction' (2013, p. 187). Nason continues,

> Indeed, they seek protection not only for themselves but for those values, practices and traditions that are at the core of Indigenous women's power

and sovereignty – concepts that have been, and remain under attack, and which strike at the core of a settler-colonial misogyny that refuses to acknowledge the ways it targets Indigenous women for destruction. (p. 187)

Nason offers up means to read Federici's 'natural resources' into Indigenous feminist activism. Leanne Betasamosake Simpson specifically describes her Nishnaabeg experience of colonialism 'as a gendered structure' adding, 'The structure is one of perpetual disappearance of Indigenous bodies for perpetual territorial acquisition, to use Patrick Wolfe's phrase' (p. 45, 2017). Together, Simpson and Nason help to frame the conceptual limits for social reproduction, each offering counter examples in Indigenous social life. Nason's 'values, practices and traditions that are at the core of Indigenous women's power and sovereignty' should not correlate with those 'changes' Federici points to in primitive accumulation that condition 'the social position of women and the production of labour-power' (2004, p. 18). As Simpson shows in what could be read as a forceful challenge to social reproduction, historically, 'Hierarchy had to be infiltrated into Indigenous constructions of family so that men were agents of heteropatriarchy and could therefore exert colonial control from within' (2017, p. 109). Colonial missions 'made Michi Saagig Nishnaabeg men into farmers and carpenters, and Michi Saagig Nishnaabeg women into wives, in the traditions of England,' Simpson explains, in terms that suggest a colonial imposition of social reproduction alongside industrial capitalism, 'meaning they engaged in unpaid labour contained inside a British-like household, ruled by their husbands.' She emphasizes, 'Michi Saagig Nishnaabeg women were transformed from being autonomous, influential, and economically powerful within our nation to being dependent, subservient, second-class citizens confined to the domestic sphere in colonial society' (2017, pp. 109–110).

While the specific circumstances for Qayqayt women would have differed, Simpson's historical details of Nishnaabeg households underscore how concepts such as social reproduction or the division of labour cannot be readily transposed in relation to histories of Indigeneity and Indigenous Traditional Knowledge. If, for instance, social reproduction is always narrowly premised on mystification (again, the term Federici uses) along a gender binary, the Anishinaabe principles of *Mino Bimaaddiziwin*, articulated in Simpson's work (2011) as well as by Indigenous feminist scholar Winona LaDuke (1999), could generate a conversation to differently conceive of the social correspondences that premise social reproduction. Winona LaDuke writes that the Indigenous relationship 'to the land and water is continuously reaffirmed through prayer, deed, and our way of becoming – *minobimaatisiiwin*, the "good life"' (1999, p. 4). Elsewhere, she suggests, 'There is no way to quantify a way of life, only a way to live it' (p. 132). Other Anishinaabe understand that, 'Each of us has been given the gift of Mino Bimaaddiziwin, the Good Life. In our language, in the way we act and think, we can choose to be, Anishinaabe, and we can choose to live, Mino Bimaadiziwin, the way our Ancestors planned for us' (Seven Generations Education Institute 2015, p. 2). As long as social reproduction *is* mystification – Federici designates, taken 'as a personal service and even a natural resource', it instantiates the reproduction of workers generationally and produces the commodity of labour-power – the concept cannot correspond to LaDuke's affirmation that 'There is no way to quantify a way of life, only a way to live it'. Labour's 'un-waged condition' does not guarantee social reproduction exceeds or escapes quantification: it remains the apparatus of capture by which women's unwaged labour still has value in capitalism.

Submitting only one or two examples from Indigenous Traditional Knowledge in dialogue with social reproduction, as I have in my reading, is by no means sufficient. But here too, delineating the limits of social reproduction, I need to be mindful of my own social reproduction and knowledge production. Again, I might turn to Simpson, who suggests her Nishnaabewin concept of 'grounded normativity' as it 'compels us to ask particular questions when we're engaging with theories from outside our own practices' (2017, p. 63). Simpson continues, 'Where does this theory come from? What is the context? How was it generated? Who generated it? ... Can I use it in an ethical and appropriate way (my ethics and theirs) given the colonial context within which scholarship and publishing take place?' (p. 63)[4]

Her crucial sequence of questions reminds me of the limitations of my own readings of 'Larrabee's poem', which, as a quotation of Larrabee and her mother reproduced in a new public context, is also the architect Graham McGarva's own poem. I am equally limited in my readings of these Indigenous feminist texts by Simpson, LaDuke, and Nason that I bring into discussion with the poem here – a position of settler readership I write about in the next section. I understand my gesture of delineating the limits of social reproduction theory as another point at which I must respect the epistemological boundaries for reading this poem in public.

Ask Again: Authorship and a Short History of the Qayqayt

Without the conventional identification of author or artist, and without standard figural representation in letters, we cannot easily come to Larrabee's poem with the understandings or techniques we have honed as readers – whether on the streets, looking at signage, or in the pages of a book. In order to carry out a straightforward reading of her poem from this position at Plaza 88, even at a passing glance, would require outside research, or the intertextual knowledge of local media coverage such as a June 2009 article that profiles the Qayqayt chief under the headline 'The Hidden Story' (Lau 2009, p. A03). While newspapers have been reporting on Larrabee and the Qayqayt since 1995, this front-page revelation calls attention to a significant part of the story: Larrabee's discovery of her family history and the history of a Nation that has been systematically erased (Bell 1995, p. A1; Pappajohn 1995, p. 1). In 1996, Larrabee became the first member of the Qayqayt in nearly 50 years (Urban Systems 2014c). Qayqayt, or the New Westminster Indian Band, had become part of the Inactive General List of Reserves in 1951. Currently, or officially, Qayqayt is one of the only Nations in Canada without a land base and one of the country's smallest Nations in terms of membership, with a total registered population of 14 (MetroVancouver 2018, p. 24).[5] As 'The Hidden Story' feature begins, 'Rhonda Larrabee still remembers the conversation with her mother as if it happened yesterday. "She told me she would tell me the story once and that we would never talk about it again ..." It was certainly an emotional talk and one that I don't think I'll ever forget.'

Like so many other street corners across so-called Canada, this site has not been ceded by treaty. Plaza 88 thus compounds the problem of site-specific art on stolen Native land. Without prior knowledge as to Larrabee's role in this publication or her relationship with the architect, Graham McGarva, readers also cannot be certain about her consent to appear on the wall. If the line constitutes a stolen text on stolen land, we reach yet another cruel dilemma.[6] In my reading, I am less interested in McGarva's role – in the decisions he makes as poet-cum-architect here that could indicate conventional authorial or artistic intentionality. Instead, I demonstrate how the readership for Plaza 88 is forced to approach this narrative in an abstract form, rendering Larrabee's 'work' – both as the actual 'author' of the poem and as the Chief who has carried out everything the text represents – into what appears to be a diminished question of aesthetics. Marie Joseph Bandura, the speaker or narrator of the 'I will tell you once' line, and her family, have endured consistent erasures and abstractions in their daily lives, which would seem to make the poem's format inconceivably cruel. I am not willing to credit McGarva with a robust critique of the politics of recognition, but I do want to reconsider the poem's illegibility as bound up in a rejection of recognition claims grounded in what Coulthard calls '"essentialist" articulations of collective identity' (p. 79). Particularly when read amongst the other poems 'published' on Plaza 88, as I will show, Larrabee's line represents a marginalized voice amongst a collection of historical texts. But even more, I want to emphasize that the poem conspicuously delineates certain limits in site-specificity because of the Qayqayt's fractured relation to land and the brutal interruptions to what Federici called the 'reproduction of their community' (Carlin and Federici 2014, para. 9). In setting up the longer historical context, I want to emphasize that within this particular scenario of site-specific art, I am offering up one specific reading. My reading is conditioned by my experiences as a white settler living for 25 years on Coast Salish territory. I only learned about the history of the Qayqayt Nation while researching local media to understand the new installations of poetry on a transit station near my parents' house. I came to Larrabee's wall in the summer of 2012, after having moved away from New Westminster to study in Minneapolis. Initially, my brother drew my attention to the site, mentioning, 'There's a lot of poetry down there,' in passing, talking about the area's recent development. Because we were both graduate students in literature, even this exchange is mired in academic inquiry. As de Leeuw and Hunt write, 'To not acknowledge these contexts risks perpetuating the idea that writing and knowledge is not produced in *places*, many of which are forged in ongoing colonial violence toward Indigenous peoples' (2018, p. 2). In their introduction to their article, 'Unsettling decolonizing geographies', these authors argue that 'an important step in writing about and envisioning practices of decolonization' is 'the acknowledgment that knowledge is always situated', underscoring, 'To not acknowledge who we are, or to leave unspecified our authorial position in relation to this paper and events unfolding all around us, is to risk perpetuating the idea that writing and knowledge is not produced *by* people who occupy specific temporal and sociocultural positions, positions often bound to or by colonialism' (pp. 2–3). The complicated relations of authorship that this poem engages require reflexivity about my own reading and its potential roles in knowledge production, and even social reproduction. As I said in introducing the poem, the literary ambiguity in Larrabee's line feels as though it is not for the reader to discover – again, to organize or orchestrate a singular 'reading' that might be 'mine' would be to ignore the epistemological politics involved in issuing such interpretation or analysis from my position.

Authorship is thus not only a problem within the publication context of Plaza 88, but an issue that extends into the writing of this chapter. My methodological approach to the poem is based in my literary training, but I try to be intentional about – if not divergent from – a standard close reading, or even the more reflexive 'active reading' that has been more common within experimental poetry and poetics. The word-by-word attention of this technique does not easily lend itself to engagements with the new public poems under my study, which are often positioned with the very aim to disrupt the everyday hustle and bustle that renders them otherwise unintelligible. In Larrabee's case, taking seriously Coulthard's critiques of the politics of recognition, meaning cannot accumulate across these 12 words in a mode that is familiar or even recognizable. If city dwellers in late capitalism conceive of urban space and the built environment as commodities themselves, understood in the global logic of post-industrial capital accumulation (Soja 1989; Jameson 1991), then the stakes for accumulation of meaning in the poem are denatured by the poem's position as a commodity that somehow adds value to the building. But my reticence as a reader here is 'worth it' – consumer pun intended – given my position as another white settler putting words together or making meaning in capitalist urban space. And yet I take responsibility for the poem as it appears in my reading; particularly because, in many other ways, it cannot actually be read. In my interdisciplinary approach, I take seriously calls from post-colonial scholars in geography like Hunt and de Leeuw, as well as Reuben Rose-Redwood. Discussing the renaming of PKOLS in Saanich, BC – the municipal park and landmark known to non-Native people as Mount Douglas and named for Sir James Douglas, renamed PKOLS in May 2013 by the Tsawout Nation – Rose-Redwood suggests that 'our conception of the horizon of toponymic politics should also extend *beyond* the politics of recognition,' and work to '*decenter* the settler-scholar as the sovereign subject of intellectual authority and political agency' (2016, pp. 201–202). While Plaza 88, at least at the moment, does not bring up immediate issues of toponymy, I take each of these statements seriously in this poem's capacities for social reproduction, recognition politics, and even in place-making as a broader urban conundrum.

In 'Ontologies of Indigeneity: The politics of embodying a concept,' Hunt writes, 'Engagement with Indigeneity involves the establishment of ontological limits around what knowledge is and is not legible,' and she argues that 'In order to be legible, Indigenous geographic knowledge must adhere to recognized forms of representation' (2014, p. 28). The problems of 'legibility' Hunt describes overlap with, and are compounded by, the illegibility I have attributed to Larrabee's poem. She 'propose[s] that these gaps in regimes of knowledge provide sites where ontological shifts are possible,' and suggests that in 'accepting the partiality of knowledge', 'as we attempt to fix its meaning, we are always at risk of just missing something' (pp. 30–31). Any effort to fill in the 'gaps' in the narrative on display at Plaza 88 – where Larrabee's poem is actually constituted by a sequence of visual gaps – would seem insufficient. Nonetheless, archival reviews of holdings at the New Westminster Museum and Archives as well as the New Westminster Public Library have been part of my research methodology. While I have drawn primarily from contemporary materials – newspapers, websites, blogs – that position the poem in urban space now, I have utilized archival photographs and newspapers as well as queried how Qayqayt history is currently being taken up within the local settler imaginary. Archaeological data presented at the New Westminster Museum show that the banks of the Fraser River – which forms the southern border of present-day New Westminster and is located across the street from the Plaza 88 site – have been the ancestral homes for Coast Salish communities for 10 000 years (*Permanent Collection* 2017). Marie Joseph Bandura was one of the last members of the group known at the time as the New Westminster Indian Band to reside on a 22-acre parcel of Qayqayt reserve land in New Westminster, near the current Kruger Paper Mill – just over a mile westward from Plaza 88 on the riverfront. Qayqayt, which translates to 'resting place', named an earlier village of around 400 people who, as Larrabee describes, were 'displaced … stricken with a smallpox epidemic. Now it's like they never existed' (Urban Systems 2014c). Reduced to less than 100, any members were assimilated into other local reserves or sent to residential schools after government seizure of the band's reserve lands in 1916. After Joseph survived the Kamloops Indian Residential School in the interior of British Columbia – approximately 200 miles from her home in New Westminster – she moved to Vancouver's Chinatown, where she married and raised her children. Larrabee and her brothers understood that their father, Art Lee, was Chinese and that Marie Joseph was French. When she was 24, assembling information for a family tree, Larrabee asked her mother for her grandparents' names. Joseph's response, 'I will tell you once, but you must never ask me again,' would set in motion a story she had kept secret for several decades.

After high school at the Kamloops residential school, Joseph and her sister Dorothy hitchhiked to Chinatown where Dorothy Joseph first found a job as a tailor and Marie worked as a waitress; Larrabee describes in the documentary film *A Tribe of One* (2003) that the young women quickly realized they could pass for 'Asian of some kind' and that by the time her mother married Art Lee, it 'was just established, we were a Chinese family'. She abided by her mother's wish 'to never ask me again', in the years after learning about their Indigenous heritage, even as, in her words, the secrets 'ate away at me, to know something and not be able to ask more questions.' A few years after Joseph passed away in 1985, Larrabee's father Johnnie Bandura, Joseph's second husband, encouraged her interest in finding out more about their heritage, urging her to 'do it' and 'make her proud' (Urban Systems 2014b). In 1994, after 13 months spent waiting and searching for documents to send to the former federal Indian Affairs and Northern Development Department, Larrabee received her Indian Status card, which made her the first documented member of the Qayqayt First Nation since 1951. Considering historical frameworks for the politics of recognition, it is crucial to clarify that Larrabee's aunt and uncle, Dorothy Isabelle and George Joseph, had been the last living members on this list, but at the time, her family did not know about the governmentally sanctioned 'Indian status' that would be required for official recognition (Bell 1995). Nearly 50 years later, once Larrabee's three brothers also gained status, the group held a council election and Larrabee was elected Chief of the Qayqayt First Nation (Urban Systems 2014b). In 2009, the Nation identified three parcels of land that would constitute a land claim – though, under pressure from concurrent claims by neighbouring groups for this unceded territory, Larrabee had emphasized in earlier interviews that 'Land wasn't the reason why I did this' (Lau 2009, p. A03).

As Larrabee describes, 'I was just so emotional when I got my card. I felt like I was part of Mom ... And that was my goal of getting my status – to feel close to where my Mom came from, her roots' (Urban Systems 2014b). But she confirmed in 2015 that the Qayqayt 'filed a claim in 2012 and the three-year limit is up this year, so we're hoping to be in negotiations soon. It's been a long time because I was granted my status in 1994' – emphasizing that she has been engaged in the process for 25 years (Verenca 2015). Metro Vancouver's *Profile of First Nations with Interests in the Region* registers that the Qayqayt are 'Not involved in treaty negotiations', but there is no information in that document about the possible sale of Musqueam Reserve No. 1, in Surrey, BC, to the Qayqayt at some point prior to November 2014 (Wall 2014, p. 24).[7] The Qayqayt's registered population in March 2018, as I noted earlier, was 14 members, making the group one of the smallest First Nations in Canada and, again, at least in governmental reports, one of the only Nations without a land base.

Larrabee has described how her 'phone rang off the hook' after gaining status (Urban Systems 2014c). She recalls how a Skxwú7mesh Úxwumixw woman working for the Katzie Nation at the time asked, 'Do you know the responsibility of you being reinstated into this band?' This query, as well as the 'overwhelm[ing]' three-hour meeting that followed, once again foreground the issue of social reproduction within Larrabee's personal history and 25 years of work for the Qayqayt. She first approached the City of New Westminster and local public post-secondary institution Douglas College, and explains that political relationships for the Qayqayt 'mushroomed out from there' (Urban Systems 2014c). Larrabee describes how gaining status '[took] years and years of work and trips to Ottawa and archives and research, with little money' (Verenca 2015) and, after retiring 18 years ago, she now dedicates her time fully to 'creating cultural awareness, especially among children' and working on the complicated, expensive process of filing a land claim with Indigenous and Northern Affairs Canada (Urban Systems 2014c). But without a land base, the Qayqayt are not eligible to work with the BC Treaty Commission, which facilitates treaty negotiations amongst First Nations and federal and provincial governments. With Urban Matters, a private consulting firm, they are developing options for a permanent reserve, as well as potential cultural and community spaces.

Colonial Legibility and the Postmodern Media of Recognition

Larrabee's 25 years of work for the Qayqayt Nation critically demonstrate just how indistinct productive labour and the reproduction of a community can become under circumstances of survival. The stakes for social reproduction in the Qayqayt community are bound up in the complications of these potential land claims and, by extension, the myriad modes and mechanisms for their recognition by state authorities such as the City of New Westminster and the provincial and federal governments. The possibility for 'reproducing' the individual members of the 14-member Nation, at this crucial juncture in their land claims or in their oft-reported position as 'Canada's only tribe without a land base', have to be constantly waged against urban capitalism and the maintenance of a particular flow of labour power, what Coulthard calls 'the surplus value afforded by cheap, Indigenous labour' (p. 12). As he explains,

> It is now generally acknowledged among historians and political economists that following the waves of colonial settlement that marked the transition between mercantile and industrial capitalism (roughly spanning the years 1860–1914, but with significant variation between geographical regions), Native labour became increasingly (although by no means entirely) superfluous to the political and economic development of the Canadian state. (p. 12)

During the swell in European settlement after the fur trade, land became the primary requirement in the establishment of the Canadian state, and Coulthard adds that within this development colonial authorities 'only secondarily' sought 'the surplus value afforded by cheap, Indigenous labour'. But the unlikely negotiation of reconciliatory politics on the walls of this building specifically engages a contemporary colonial power struggle that Coulthard calls 'the core theoretical intervention' of *Red Skin White Masks*: amongst his revisions to Marx's theory of primitive accumulation, he registers a change in 'colonial relations of power' in Canada. Historically, these power relations were 'reproduced primarily through overtly coercive means', but Coulthard diverges from Marx – who saw violence at the root of dispossession and its subsequent accumulation – to argue that today colonial power comes through 'the asymmetrical exchange of mediated forms of state recognition and accommodation' (p. 15).

I am interested in how Coulthard's characterization of an 'asymmetrical exchange' can be read, and thus also approached formally or experienced physically, across the surfaces and references of Plaza 88. Indeed, 'asymmetry' would seem to be a distinguishing feature of Larrabee's mural, marked as it is by an awkward checkered indentation of the first fractured word, 'I', that is nonetheless balanced at the same time by a poetic 'line-break' of the same size at bottom right. But Plaza 88's other poem, written by the building's chief architect Graham McGarva, reads, 'In making Canada, a tented canopy set upon a hill,' forcing a charged aggregation, at this seemingly inconspicuous site, of conflicting monumentalities and histories fused quite literally just around the corner from one another (see Figure 3.3).

Particularly when read in conjunction with the other primary poem on the building, the decision to obscure Larrabee's message appears as an extension of colonial and gendered violence upon the text's Indigenous author and subjects. In stark contrast to Larrabee's poem, McGarva's

Figure 3.3 Southern elevation: Plaza 88. Poem by Graham McGarva. New Westminster BC. (July 2018, photo by Meghan Armstrong).

own poem is rendered in a large, uniform typeface, the capitalized letters composed of stainless steel cut into a sans-serif font, easily seen even from heavy traffic on Columbia Street.[8] Anecdotally, when I mention the building to anyone who might have seen it in New Westminster, these readers are quick to recall the clearly typed colonial poem, while only vaguely remembering Larrabee's mural. But even at this more informal register, I want to avoid displacing the recognition politics bound up in each opportunity to read or just to 'notice' Indigeneity here. The reasons why settler readers cannot see Larrabee's poem remain paramount.

Returning to the point of illegibility, the visual asymmetry here underscores how the unreadable poem forces these processes of recognition upon us, as readers, to recognize and maybe reject. Together, the two poems provide their readers with an illusory occasion to observe, if not contemplate, the 'forms of state recognition and accommodation' that permeate Canadian everyday life and show up on city street corners in ever-changing configurations. Coulthard (2014) quotes Taiaiake Alfred, who argues that 'under these "postmodern" imperial conditions', 'oppression has become increasingly invisible; [it is] no longer constituted in conventional terms of military occupation, onerous taxation burdens, blatant land thefts, etc.', but instead occurs through 'a fluid confluence of politics, economics, psychology and culture' (Alfred as cited in Coulthard 2014, p. 48). I cannot help but read this 'fluid confluence' alongside Cindi Katz's feminist articulation of social reproduction as 'the fleshy, messy, and indeterminate stuff of everyday life [that] is also a set of structured practices that unfold in dialectical relation with production, with which it is mutually constituted and in tension' (2001, p. 711). Obviously, Alfred's characterization of oppression is not interchangeable with Katz's concept of social reproduction; rather, I see how 'postmodern imperial conditions' are increasingly contingent upon social reproduction, especially at the level of the 'structured practices' of the state.

The image of a tent mobilized 'in making Canada', to return to McGarva's own poem, obviously represents the dwellings of the colonial parties who would displace Indigenous residents from their land. This tent does not exactly veil the colonial logic here; from its position on a fifteen-foot panel of concrete, McGarva's poem presides over five lanes of traffic and three train tracks, poised at one of the busiest intersections in the city. This section of Columbia Street is designated as a Major Road Network and positioned between a Provincial Highway and an Arterial Route that account for 36 200 and 46 500 vehicles per day, respectively, in average weekday travel volumes (Urban Systems 2014a, p. 26). The potential readership at this particular corner is supplemented by the 7300 average vehicles per day that pass by Larrabee's elevation on Eighth Avenue, which has been designated a City Collector road (Urban Systems 2014a,

p. 26), and, of course, by as many as 27 900 weekday Skytrain passengers who have used New Westminster Station in recent years (TransLink 2016, p. 7). At these volumes, 'In', as opening preposition, reads almost as an invitation for every reader, in their simultaneous capacities as drivers, passengers, pedestrians, or transit users, to participate in this national process of 'making'. Reflexively, McGarva's poem reminds these readers that they are already part of this development.

In an attempt at summation, McGarva has suggested, 'There's a whole bunch of literary allusions ... It's bold. It might not be to everyone's tastes, but there's a proud story to this building' (Granger 2012). An earlier article in *The Record* newspaper explains that Plaza 88 explicitly manages several narrative elements in order to 'tell the story of New Westminster, which *includes the land of the First Nations and the settlement of New Westminster*,' ('History Retold at Plaza 88' 2011, p. A05, emphasis added). The article's brief description of the 'inclusion' of Indigenous subjects draws attention to how McGarva employs literary techniques of counternarrative – bringing the friction of monumentality and its critiques into direct juxtaposition. It is worth briefly unpacking the nuances of this description of the building; I would argue that even the quick syntax of this statement reproduces the imbalance between narrative and counternarrative, colonizer and colonized. Though this short description accurately reflects historical chronology, 'the settlement of New Westminster', which describes European establishment of the city in the decades after the arrival of European explorers in 1776, comes after the 'land of the First Nations', the dominant cultural narrative of colonialism guarantees the primacy of 'settlement' as a framework within which Indigenous subjects must be 'included'. It is assumed that the readers of this article, as well as the 'readers of the building' will have to 'include' 'the land of the First Nations' within their prejudiced perspectives on local history. But while the verb 'includes' posits the 'land of the First Nations' as an element within 'the story, of New Westminster', it can also denote a sense of permission that precludes any possibility for the 'included' element to be part of a whole. Perceiving incorporation in this ancillary way represents another displacement of the counternarratives that appear on the building. To borrow briefly from Coulthard's critique of Nancy Fraser's status model, we might consider how even within an otherwise banal report on Plaza 88's construction progress, colonial dominance is so entrenched in language that we can see evidence of 'the problematic background assumption that the settler state constitutes a legitimate framework within which Indigenous people might be more justly included, or from which they could be further excluded' (p. 38).

In another conversation, McGarva explains that, 'the tented canopy structure in the plaza with all its "weird angles" is a homage to the tents of a century and a half ago' (Granger 2012).[9] Apparently he 'saw the building as a raw canvas that could tell many stories.' After 'including the land of the First Nations', this troubling perception of the site as a 'blank canvas' obviously furthers colonial connotations, but I want to focus instead upon how Plaza 88's 'tented canopy' rooves allow for McGarva to arrange poetic repetition across the media of architecture and text – all to promote the dubious symbol of the 'tent' (see Figure 3.4). Structurally, the 'weird angles' seem to be more immediate consequences of Plaza 88's relationship to the existing Skytrain station on site. As the VIA website's project details describe (2018), 'Plaza 88 represents the regeneration of five city blocks that wrap around the New Westminster Skytrain Station, located between the City's historic core and its riverfront.' Entering the plaza from street level, a visitor would be more likely to experience the oppressive shelter of the concrete train tracks that bisect the centre of the plaza than the poetic arrangement of the tented rooves that McGarva describes. Seeking cover here has been, in my experience, a negotiation of incessant coastal rainfall against the 'weird angles' of the roof three floors above, which cannot seem to quite cover all the bases below. Materially, the rooves offer up a syncopation that extends into the semiotics of the line of poetry. If we test the weird angles of the rooftops on Plaza 88 as contemporary stand-ins

Figure 3.4 'Tented roof', Plaza 88 courtyard interior, upper level. (July 2018, photo by Meghan Armstrong).

for the colonial 'tented canopy set upon a hill', 'making Canada' can be an ongoing process, the particularity of the gerund signalling futurity. But this reading is a benevolent one, bursting the limitations that have already been set by historical allusion. The new tents here, precisely as stand-ins, serve as more monuments to the brutal foundations of Canadian colonialism, still on stolen land.

After all, we are looking quite literally at the poetic wrapping paper for a shopping mall. The scrambled, denatured monument to Qayqayt history emblazoned across a high-volume transit route would seem emblematic of, in Federici's paraphrasing of ecofeminist critiques, 'the way in which capitalism has freely appropriated women's labour and the way in which it has appropriated the wealth of nature: lands, seas, forests, all treated as free resources to be used, destroyed, even exhausted without any thought of the social and ecological cost involved' (Carlin and Federici 2014, para. 9). In its fragmented quotation of Larrabee – the female chief of the 14-person Qayqayt Nation – can the poem be read as 'appropriation' of her 25 years of work and leadership that has, along with everything else, involved calling attention to 'the wealth of nature' seized from this place's original residents? Every recent possibility for the Qayqayt Nation's social reproduction has only been realized through Larrabee's efforts – securing the potential for an 'again' exactly where the poem delimits 'never'. Conversely, across its winking checkerboard of greyscale panelling, it might become possible for us to see these fissures, bound up in the impossibility of reading a story that both should and should not be told here, in these terms and through channels of urban capitalist development. These are not merely issues of documentation or historical representation, nor can the poem be considered just a problem of commodification – or even mallification. What if Larrabee's poem had been rendered in the same font as McGarva's 'making Canada' line, at its legible size and scale? Would it make sense, or does the language of the poem itself – a statement that sets up 'I will tell you once, but you must never ask me again' without resolution – foreclose upon a recognizably monumental exposition? I am tempted to look at a statement Larrabee offered at a more directly state-sanctioned site half a mile from Plaza 88: École Qayqayt Elementary, responding to the new school's name at its groundbreaking in April 2013:

> For a long time, many, many years, the Qayqayt Nation was not recognized in New Westminster – for all kinds of reasons, which would take me a couple of hours to explain. But to have the legacy of our ancestors, our family, our traditions, to have them all being carried out today is what's important. And we're so excited about having this school named after our First Nation, I just can't tell you. We're just over the moon about it. (The Royal City Record 2013)

It is not for me to recount 'all kinds of reasons' why the opening of the school opens up different questions about the politics of recognition. Moving momentarily from the poem at Plaza 88 to Qayqayt Elementary is, rather, an attempt to replace us on an ostensibly more stable monumental ground. But obviously, after all we have seen, the naming of this site is in no way clearer than the processes of identification or social reproduction at work in the poem. It is merely more familiar – though again, I do not want to diminish the school's significance, especially as a student who studied for 13 years on top of the hill and never learned about the Qayqayt until I moved away and started work on my third university degree.

Larrabee has long been associated with New Westminster's World Poetry Reading Series, performing official welcomes and organizing events. I have not found any record of her speaking publicly about 'her poem' on Plaza 88. But she speaks frequently about her commitment to children and public education; she has advocated for the history of residential schools to be taught at the elementary level (Verenca 2015). As well, Larrabee often foregrounds her own family's roles in operations of the Qayqayt Nation, pointing out that 'They're all extremely involved. They've all been involved in our [land] claim process and I can't do anything without consensus from the group.' She adds that, 'everyone's excited about the possibility that we might get a land base and be able to have a community like there was before. It's been really good because now they say they know who they are. They know where they came from' (Larrabee as cited in Verenca 2015). Despite gathering more details about the central positions of children and families in Larrabee's work, I hesitate to impose social reproduction as its characterization. As I mentioned earlier, concepts like those encompassed in Anishinaabe *Mino Bimaaddiziwin* can call attention to the idiosyncrasies of urban social reproduction that might only get its practitioners to the point of (more) mystification – and alienation. Taking seriously the limitations of my own reading, I am still unwilling, and, if I am to trust at least in my own tentative arguments, actually unable to make sense of the grey and white mural that brings Chief Rhonda Larrabee's line to an intersection in New Westminster, on Coast Salish territories. There is no way to track all that has been lost here. But I am grateful for the opportunities reading offers up – even if only in this instance. I follow poet Cecily Nicholson out, writing in *From the Poplars*, another history for New Westminster:

> I have circled the same spot over and over
> walls rise and fall to better walls
> mortgaging future conditional
> pledge of properties

outcome larger
drain
toward doubt

(2014, p. 64)[10]

Acknowledgements

First and foremost, this doctoral research is the result of conversations, as well as other concrete research and writing efforts, carried out in cooperation with family and friends who reside in New Westminster or nearby on Coast Salish territory. Part of this research has also been supported by the University of Minnesota Graduate Research Partnership Program and the Canada Social Sciences and Humanities Research Council.

Notes

1 Plaza 88 was reopened as Shops at New West in November 2012. I retain the site's original name in my research for clarity in reviewing the print materials that were published at the time of its construction, all of which use the Plaza 88 name.

2 Douglas Crimp in *On the Museum's Ruins* attributes site-specificity to minimalist sculptors who 'launched an attack on the prestige of both artist and artwork, granting that prestige instead', he describes,

> to the situated spectator, whose self-conscious perception of the minimal object in relation to the site of its installation produced the work's meaning ... This condition of reception, in which meaning is made a function of the work's relationship to the site of exhibition, came to be known as site-specificity ... (Crimp 1993, p. 17)

'Once the works are erected in a public space,' Crimp writes, 'they become other people's concerns' (p. 165).

3 For this concept I draw from critical essays by the North American L = A = N = G = U = A = G = E poets, publishing since the late 1960s. For example, see Ron Silliman, who writes in *The New Sentence*, 'Words not only find themselves attached to commodities, they *become* commodities and, as such, take on the "mystical" and "mysterious character" Marx identified as the commodity fetish' (1977, p. 8).

4 Simpson's series of questions reads, in full,

> I think grounded normativity in its Nishnaabewin formation compels us to ask particular questions when we're engaging with theories from outside our own practices. Before I use work by writers, scholars, and

artists outside of the Nishnaabeg nation in my own writing and thinking, I ask myself the same series of questions. Where does this theory
come from? What is the context? How was it generated? Who generated it? What was their relationship to community and the dominant
power structures? What is my relationship to the theorist or their community or the context the theory was generated within? How is it useful
within the context of my own people? Do we have a similar concept or
theory? Can I use it in an ethical and appropriate way (my ethics and
theirs) given the colonial context within which scholarship and publishing take place? What are the implications of citation, and do I have
consent to take this intellectual thought and labour from a community I
am not a part of? Does this engagement replicate anti-Blackness? Colonialism? Heteropatriarchy? Transphobia? This critical process, I think,
is a process many Indigenous academics already do naturally, and the
answers are not easy, nor will they be the same for everyone. (p. 63).

5   Later in this chapter, I explain the discrepancies in information about the
status of potential land claims for the Qayqayt Nation. (See Note 7).

6   In these interviews from local media, McGarva's relationship to Larrabee
(and her quotation) is not clear.

7   Notes from a University Endowment Lands (UEL) Project Workshop (Wall
2014) dated 13 November 2014 indicate the sale of Musqueam Reserve
No. 1, in Surrey, BC, to Qayqayt prior to 2014. Again, this potential claim
is not listed in municipal or provincial governmental documents associated
with the Qayqayt, and I have not yet been able to confirm its validity.

8   The dominant voices still have the most power and legibility here; I read
McGarva's 'making Canada' poem as secondary or supportive to Larrabee
in some small effort to shift this balance.

9   In its historical photo collection, the New Westminster Public Library has
several images of tents categorized under 'Camps and Camping'. These
'camps', which are identified as *either* settler camps or Indigenous camps
in the limited notes on the images, are useful in contextualizing how this
symbolic element of McGarva's poem swings between the two cultural
groups represented in his line; the generality of the library's subject head-
ing underscores the historical ambiguity of this scene. An original photo
was captioned 'The Pleasure Grounds', and the campsite is otherwise situated near 'the Condensed Milk Factory', 'East of first Fraser R. bridge'.
Historian Archie Miller suggests the possibility of the tents being located
at the 'Indian Camp which was on the east side of the New Westminster
Bridge.' With these two possible identifications in mind, the symbolism of
McGarva's tents is irreconcilable.

10 Governor General Award-winning poet Cecily Nicholson's *From the
Poplars* (2014), which won a BC Book Prize in 2015, documents histories
of New Westminster and the Qayqayt Nation collected around Poplar
Island, a depopulated island in the Fraser River that was once the site of
a 27-acre reserve for the Qayqayt as well as a traditional burial ground
(McKenna-McBride Royal Commission 1916, p. 685).

References

Bell, S. (1995). One person band has big plans. *Vancouver Sun* (4 October), p. A1.

Carlin, M. and Federici, S. (2014). The exploitation of women, social reproduction, and the struggle against global capital. *Theory & Event* 17 (3): Project MUSE, muse.jhu.edu/article/553382.

Coulthard, G. (2014). *Red Skin White Masks: Rejecting the Colonial Politics of Recognition*. Minneapolis, MN: University of Minnesota Press.

Crimp, D. (1993). *On the Museum's Ruins*. Cambridge, MA: MIT Press.

de Leeuw, S. and Hunt, S. (2018). Unsettling decolonizing geographies. *Geography Compass* 12. doi:10.1111/gec3.12376.

Dean, A. (2015). *Remembering Vancouver's Disappeared Women: Settler Colonialism and the Difficulty of Inheritance*. Toronto: University of Toronto Press.

Eng, M. (2016). *Mercenary English*. Second edition. Vancouver, BC: Mercenary Press. https://mercenaryenglish.files.wordpress.com/2016/09/mercenary-english.pdf (accessed 1 December 2016).

Eng, M. (2017). *Prison Industrial Complex Explodes*. Vancouver, BC: Talon Press.

Federici, S. (2004). *Caliban and the Witch: Women, the Body, and Primitive Accumulation*. Brooklyn, NY: Autonomedia.

Granger, G. (2012). Poetry, history on New Westminster's Plaza 88 façades. *The New Westminster NewsLeader* (6 July), p. A1.

History Retold at Plaza 88 (2011). *New Westminster Record* (24 June), p. A05.

Hunt, S. (2014). Ontologies of indigeneity: The politics of embodying a concept. *Cultural Geographies* 21 (1): 27–32.

Jameson, F. (1991). *Postmodernism, or, the Cultural Logic of Late Capitalism*. Durham, NC: Duke University Press.

Katz, C. (2001). Vagabond capitalism and the necessity of social reproduction. *Antipode* 33 (4): 709–728.

LaDuke, W. (1999). *All Our Relations: Native Struggles for Land and Life*. Boston, MA: South End Press.

Lau, A. (2009). The hidden story. *New Westminster Record* (6 June), p. A03.

McKenna-McBride Royal Commission (1916). Minutes of decision – New Westminster agency. Final report – New Westminster agency. *Union of BC Indian Chiefs*. www.ubcic.bc.ca/mckenna_mcbride_royal_commission (accessed 1 September 2017).

MetroVancouver (2018). Profile of First Nations with interests in the region. *MetroVancouver Aboriginal Relations*. http://www.metrovancouver.org/services/first-nation-relations/AboriginalPublications/ProfileOfFirstNations.pdf (accessed 10 March 2018).

Murakami, S. (2008). *The Invisibility Exhibit*. Vancouver, BC: Talon Press.

Nason, D. (2013). We hold our hands up: On indigenous women's love and resistance. In: *The Winter We Danced: Voices from the Past, the Future, and the Idle No More Movement* (ed. Kino-nda-niimi Collective), 186–189. Winnipeg, MB: ARP Press.

Nicholson, C. (2014). *From the Poplars.* Vancouver, BC: Talon Press.

Pappajohn, L. (1995). Search takes her home…. *Royal City Record NOW* (4 June), p. 1.

Permanent Collection [Exhibition] (2017). New Westminster Museum and Archives, New Westminster, BC. Ongoing.

Rose-Redwood, R. (2016). Reclaim, rename, reoccupy: Decolonizing place and the renaming of PKOLS. *ACME: An International Journal for Critical Geographies* 15 (1): 187–206.

Rosler, M. (2013). *Culture Class.* Berlin: Sternberg.

The Royal City Record (2013). Qayqayt elementary groundbreaking ceremony. *YouTube.* www.youtube.com/watch?v=JVMLnokphgY (accessed 3 April 2018).

Seven Generations Education Institute (2015). What is Mino-bimaadiziwin? www.7generations.org/?page_id=2822 (accessed 1 September 2018).

Silliman, R. (1977). *The New Sentence.* New York: Roof.

Simpson, L. (2011). *Dancing on Our Turtle's Back: Stories of Nishnaabeg Re-creation, Resurgence, and a New Emergence.* Winnipeg, MB: ARP Press.

Simpson, L. (2017). *As We Have Always Done: Indigenous Freedom through Radical Resistance.* Minneapolis, MN: Minnesota University Press.

Soja, E. (1989). *Postmodern Geographies: The Reassertion of Space in Critical Social Theory.* London: Verso.

Statistics Canada (2016). Census profile, 2016 census: New Westminster, City [Census subdivision], British Columbia and Greater Vancouver, Regional District [Census division], British Columbia. www12.statcan.gc.ca/census-recensement/2016/dp-pd/prof/details/page.cfm?Lang=E&Geo1=CSD&Code1=5915029&Geo2=CD&Code2=5915&Data=Count&SearchText=new%20westminster&SearchType=Begins&SearchPR=01&B1=All&TABID=1 (accessed 3 September 2018).

TransLink (2016). Appendix E: Skytrain and West Coast express line summaries. *TransLink: 2016 Transit Service Performance Review.* www.translink. ca/-/media/Documents/plans_and_projects/managing_the_transit_network/2016-TSPR/2016-TSPR-Appendix-E-Rail-Line-Summaries.pdf?la=e n&hash=25BA1DD27359A6A9F8284E41445CD57CB29826DB (accessed 3 September 2018).

Tribe of One (2003). [DVD] Directed by Eunhee Cha. Montreal: National Film Board of Canada.

Urban Systems (2014a). New Westminster master transportation plan. https:// www.newwestcity.ca/council_minutes/0616_14/2014-06-11%20New%20 West%20Master%20Transportation%20Plan%20DRAFT.pdf (accessed 3 September 2018).

Urban Systems (2014b). Reclaiming roots: A mother's anguish transforms a painful history [Part 1], 13 February. http://urbanblair.wpengine.com/ reclaiming-roots-a-mothers-anguish-transforms-a-painful-history (accessed 3 April 2018).

Urban Systems (2014c). Reclaiming roots: Unearthing the lost history of Qayqayt First Nation [Part 2], 12 March. http://urbanblair.wpengine.com/reclaiming-roots-unearthing-the-lost-history-of-qayqayt-first-nation (3 April 2018).

Verenca, T. (2015). You can't even imagine how they treated us. *New West-minster Record* (9 June). www.newwestrecord.ca/news/you-can-t-even-imagine-how-they-treated-us-1.1962618 (accessed 15 September 2016).

VIA (2018). *Plaza 88 Mixed-Use.* www.via-architecture.com/project/plaza-88 (accessed 3 April 2016).

Wall, D. (2014). Notes from November 13, 2014 University Endowment Lands (UEL) Project Workshop. http://docs.openinfo.gov.bc.ca/d19266715a_response_package_csc-2015-00005.pdf (accessed 1 September 2018).

4

Gender in Resistance

Emotion, Affective Labour, and Social Reproduction in Athens

Mantha Katsikana (York University)

Introduction

Amidst the 'Greek crisis' of austerity, Athens, a city governed through revanchist neoliberal policies, is facing the impact of intensifying social antagonisms of a collapsing welfare state, unemployment, and violence. In this context, the resistance movements that flourish in Athens embody alternative imaginations of collectivity, solidarity, and egalitarianism against the racist, classist, and other exclusionary practices of the neoliberal city. Despite the fact that issues of sexism, homophobia, and gender oppression within the Greek anarchist/anti-authoritarian movement are widely known, much academic research has focused primarily on the protest practices organized in downtown Athens.[1]

While its gendered 'micropolitics of solidarity' (Mott 2018) remain undertheorized,[2] as a result, the body of scholarship on this movement fails to engage with gender, emotional labour, or social reproduction, focusing instead on a genderless (male) crowd, on direct action as the predominant mode of resistance, and on the assumption that the experience of resistance varies only in terms of class and race.

In this chapter, I examine the relations between the city, social movements, and the commons through a case study of the anti-authoritarian/anarchist commons in Athens, in order to illustrate how the affective and collective labour of resistance by groups within these movements

A Feminist Urban Theory for our Time: Rethinking Social Reproduction and the Urban, First Edition. Edited by Linda Peake, Elsa Koleth, Gökbörü Sarp Tanyildiz, Rajyashree N. Reddy & darren patrick/dp.

contributes to the city's social reproduction. I draw from the work of both Greek and other feminist scholars on the impact of the neoliberal governance of austerity in Greece and on the gender dynamics of social protest and resistance. I argue that the affective labour of social reproduction and other practices of collective (and self-) care and well-being within activist communities are gendered and vital to the social reproduction of resistance in a neoliberalizing city during austerity, creating safe spaces against dispossession, state violence, and police brutality as well as against a patriarchal oppression rooted in Greek society. In this context, I explore the ways in which the gender dynamics of protest and resistance impact upon how female and queer activists mobilize, experience, and reclaim urban collective and public spaces in the city of Athens.

I conducted the research for this chapter between 2014 and 2017 in Athens, although some of the experiences and incidents discussed in the interviews date back to 2007, due to the importance of the 2008 youth uprising[3] and its impact on the formation of commons for anti-authoritarian communities in Greece. During the research process, I conducted 36 semi-structured interviews with individuals of various gender identifications, all of whom were either already members of groups and collectives or who had been frequently participating in some anti-authoritarian activities and self-identified as anti-authoritarians, anarchists, or autonomists.[4] In addition to the interviews, I practised participatory observation and content analysis (of zines, brochures, posters, stickers, books, and music). Practising reflexivity, I considered my own standpoint affected my relationship with the individuals I interviewed as well as others in the anti-authoritarian community. Given my personal experiences of oppression and sexism in this community, my attempt to articulate this experience in an academic context, enables a more robust interpretation of the Greek anti-authoritarian struggle and commoning under the feminist principle that the personal is political.

Protest and Resistance in Athens

Athens has emerged as the geographical and political focus of the Greek crisis, a crisis that while having a significantly gendered character affecting women as embodied subjects (Vaiou 2014, 2016), has not resulted in the gendered aspects of the crisis being seen by analysts as integral to the 'main problem' (Vaiou 2014, p. 534). As noted by Greek scholars[5] the normalization of hate discourses and the violence generated by police brutality and the rise of the far-right during the crisis have normalized and intensified sexism, misogyny, and violence against women, both domestic and public (Vaiou 2014). Anna Carastathis (2015) further argues that the violence and oppression generated in the neoliberal city during the crisis creates an 'affective economy of hostility' (p. 75) that articulates racialized and gendered modes of belonging as being estranged from the political community. Some bodies 'are rendered vulnerable, precarious – even socially dead; [while] others assert an entitled relation to national space – even as they may be economically disentitled by austerity measures' (Carastathis 2015, p. 75). Thus 'survival emerges as a politically saturated struggle' (Athanasiou and Kolocotroni 2018, p. 270). At the same time, while the accounts of resistance during the 'crisis' praise collective action, pedagogies, and possibilities of alternative futures that such mobilization and political engagement entails, Vaiou (2014, p. 536) stresses that such accounts do not refer to which bodies devote their 'time and passion' to keeping resistance going.

During the ongoing crisis period, certain episodes of extreme violence appeared as crucial, not only for the eruption of resistance movements and the politicization of public space but also for an extremely violent response from the state and the police, especially in the form of 'scapegoating' (Markantonatou 2015, p. 205). Anti-authoritarian protests and commoning were often targeted in political and media discourses as the part of an alleged 'ghettoization' of the inner-city core,[6] leading to intensifying zero tolerance policies, raids, and police brutality targeting anti-authoritarians and migrants.

During the peak of a wave of police brutality, 15-year old student, Alexandros Grigoropoulos, was shot by a policeman in the downtown area of Exarchia in December 2008. This murder fuelled a violent youth uprising that gave birth to new forms of protest and organization throughout the country, from large demonstrations to neighbourhood assemblies (Markantonatou 2015). Following this uprising, Athens mayor, George Kaminis, introduced his development agenda for downtown Athens, ironically appropriating the right to the city to promote a revanchist gentrification vision. During the same period, attacks, 'complementary' to the city's gentrification agendas, were undertaken by far-right groups,[7] in the central neighbourhoods of Athens populated mainly by working-class immigrant populations. This tolerance of far-right attacks against anti-authoritarians, immigrants, and vulnerable groups as part of policing the crisis peaked with the killing of the activist Pavlos Fyssas, which eventually resulted in the arrest of *Golden Dawn* members.[8] In the years following these episodes, zero tolerance policies in accordance with the gentrification and securitization agendas manifested in a nationwide series of raids targeting occupied buildings and anarchist squats, which led both to the arrests of anarchists and activists and secured the spatial politics of neoliberalism (Dalakoglou 2013; Apoifis 2017). Interventions that have been attending to the needs of vulnerable social groups performed by anti-authoritarians through social centres and refugee shelters in (previously abandoned) occupied buildings are not only criminalized as illegal but also as attracting unwanted land uses and flows of users that do not align with the creative agendas promoted for the touristic rejuvenation of downtown Athens.[9]

In this context, the consciousness raised by massive protests against state, police, and far-right violence led to the formation of new ways of commoning and participation. In the years following the 2008 uprising, the anarchist/anti-authoritarian movement acquired a large following with women and queer folk joining protests and political groups under the hope that being 'against all forms of power' guaranteed a safe egalitarian space for gender expression, as queer and anti-patriarchal agendas were often emphasized by groups in order to attract participation. However, the patriarchal foundations of the Greek society that have enabled public harassment, violence, and rape culture, do not stop at activist cultures and communities, revealing the necessity of the affective labour performed by anti-authoritarian women and queer folk. While the deaths of Grigoropoulos and Fyssas, both of whom were Greek and male, were enough to fuel the revolt and provoke a state response to racist and violent episodes, the violent attacks against migrants, women, and LGBTQ individuals by neo-Nazis or police and state violence, remained unspoken and attracted very little solidarity. While Konstantina Kuneva, a Bulgarian immigrant and active trade unionist,[10] who endured an acid attack in 2008, became a symbol of resistance, this was on the basis of her syndicalism and not her gender. The 17 HIV+ women – among them migrants, substance users, homeless, and trafficking victims – arrested in May 2012 under the pretence of protecting the 'Greek family'[11] – were only supported by certain feminist groups and solidarity initiatives.

The unwillingness to genuinely address problematic masculinism within the anti-authoritarian community implies that sexism, LGBTQ-phobia, and gender violence are only situated in the Greek household, in the spaces of feminized labour and exploitation, in the public spaces where women and queers are in danger of a 'Greek boy' attack,[12] and not in the spaces of resistance. The masculinist activist cultures that have dominated the public spaces of resistance during the crisis are thus part of the wider capitalist patriarchal oppression and violence spread within Greek society. As Sears (2017, p. 188) notes, it is the very repertoires of resistance that use direct action to confront the power of the state, the employer, or the police, that are central also to the perpetuation of rape culture that is reproduced 'in activist spaces because of the usefulness of masculine aggression in militancy.' This oppression normalizes gender violence and dictates who is able to mobilize in public and under what circumstances and whose voice can be heard and to what extent, marking one's 'proper place' in the social spaces of the city.

Feminist Social Reproduction in the Context of Urban Activism

The social reproduction of the urban entails both material and affective, connective, and emotional labour performed within the context of everyday life and the creation, management, and maintenance of urban commons constituted by practices of resistance and alternative forms of resilience (Kousis 2017) in the neoliberal city. The practices of social reproduction engaged in by the anti-authoritarian/anarchist movement in Athens are an integral part of a 'post-capitalist urban commons' (Tsavdaroglou 2016). I follow Jeppesen et al. (2014) in approaching the complex configurations of people, spaces, and social networks that comprise the anarchist commons as being characterized by: collective autonomy (organizing distinctly from NGOs, political parties, social services, unions, or other top-down organizations); self-determination; and self-organization via horizontal processes such as consensus decision-making.[13] Central to this commoning is the element of organizing linked to 'everyday resistance', which is practised through the interaction of spatially organized activities, social relations, and identities (Johansson and Vinthagen 2016).

In the context of everyday resistance and organizing, social reproduction emerges as a set of material and non-material practices built on the basis of feminized affective, emotional, and connective labour. Following Oksala's (2016) articulation,[14] affective labour can be seen as 'the labor of human contact and interaction, which involves the production and manipulation of affects' whose 'products [are] relationships and emotional responses' and thus cannot be always monetized but rather must be counted as 'positives externalities' (p. 292). The social reproduction of resistance in toxic spaces of masculinist activism entails the affective labour of creating social networks, attachments, and desires that form the bases for alternative modes of commoning and living, as well as the labour of confronting and resisting oppression and gender violence coming from different sources, both within the spaces of community and of the city. This affective labour, that challenges the heteronormative masculinist structuring of urban space through the formation of and constant demand for truly egalitarian safe spaces, could not be performed without the emotional labour devoted to healing from the gendered violence of both the state and masculinist behaviours. Fundamental to the social reproduction of resistance in the city, and integral to affective labour, is connective labour, an often invisible or under-estimated labour performed by women and queer folk as a 'glue' that creates the necessary conditions for 'inclusive, lateral, non-hierarchical learning or even consciousness-raising public spaces' (Boler et al. 2014, p. 444). This set of practices enables the maintenance and the 'spreading' of resistance, through collective and individual acts of care, emotional support, production of knowledge, and organizing. As a set of practices that both ensure and reproduce resistance in the city, the labours of social reproduction are deeply political, insofar as they encompass the formation of subjectivity, identity, kinship, and community, as well as the ethics by which individuals live.

The valuable 'positive externalities' of affective labour are often overshadowed by 'activist burnout' (Kennelly 2014; Chen and Gorski 2015), which entails self-doubt, physical and mental health issues, such as depression and post-traumatic stress disorders, and develops not only from the everyday-ness of activist organizing, but also from its non-recognition within masculinist activist cultures of 'selflessness' (Craddock 2019, p. 138). The emotional cost of such toxic spaces within anarchist commons is that women not only do not want to engage in the full range of the community's activities, but they also leave the movement and, ultimately there is an under-representation of women and feminized subjects.[15] This cycle of burnout reproduces further 'male-dominated organizational models and cultures ... discriminatory practices, explicit sexist expressions, and sexual harassment' (Bhattacharjya et al. 2013, p. 286).

The socially reproductive labour of women and queer folk performed through anti-authoritarian commoning has been able to create the conditions necessary for the anti-authoritarian movement to proceed with its socially reproductive and transformative struggles in a neoliberalizing city, by starting this transformation in its own core. Thus, the struggle for urban social justice, for reclaiming public spaces of everyday life, for creating grassroots responses to the collapsing welfare state and neoliberal revanchist policies of gentrification, forced evictions, and zero tolerance policy, could not be performed without the vital political feminist praxis within the city, within the neighbourhood, and within the local commons that enables the social reproduction of resistance and of alternate forms of living.

Placing Social Reproduction in the Anti-authoritarian/ Anarchist Commons

While anti-authoritarian/anarchist commons spread throughout the city of Athens, a significant number of the organizing activities and spaces are situated in Exarchia, a downtown neighbourhood known as the epicentre of alternative youth cultures and radical political activity since the days of the youth uprising (which occupied and organized from within the Athens Polytechnic situated in the area) that overthrew the military Junta in 1973.

In the years since, Exarchia has attracted a diverse body of residents and regulars due to the concentration of alternative entertainment, occupied buildings turned into social centres and shelters, and cultural spaces for underground youth cultures in the area, as well as the presence of the Athens Polytechnic, an historic university building that serves also as a bastion of political organizing and a refuge in itself. The area is heavily policed and is often subject to zero tolerance policies and subject to discourses of criminality both from the national and municipal police. In short, the anarchist commons are under a constant threat of raids and forced evictions, practices that usually take place in waves and which have intensified across the country since the 2008 uprising, but also more recently through the forced evictions of refugees from occupied buildings run by anti-authoritarian activist groups as shelters/social centres.

Usually united in alliance under urgent circumstances, such as the 2008 uprising, the various and diverse collectives and individuals found under the anti-authoritarian umbrella (Apoifis 2017) form a kaleidoscope of radical activity that fosters a variety of groups that differ in terms of their chosen organizing practices, ideological roots, attitude towards violence, inclusion of other radical agendas, and socio-temporal character (including those that are ephemeral, long-term, spanning as a network across the country, or place-based focused on direct action). Simultaneously, within the context of the crisis, the needs of everyday life have been translated into small-scale place-based initiatives such as neighbourhood assemblies, in which anti-authoritarians, anarchists, and leftists often participate as individuals and which can be found in working- and middle-class neighbourhoods. These activities have led to the spreading of anarchist ideas and ethos into politically diverse groups of people, organizing around neighbourhood issues, in localized commons and solidarity networks (Vaiou and Kalandides 2017).

Along with protests and cultural events bearing political messages, commoning entails a variety of socio-spatial practices that are usually shaped around occupation and urban squatting. Occupation can take the form of a more ephemeral direct action, often as a form of protest and organizing, aiming to disturb the order of the neoliberal city and its tourist attractions. Urban squatting, on the other hand, is usually practised as a long-term ongoing response catering for the needs of both activist and vulnerable groups. The creation of social centres, shelters, communal forms of dwelling, and organizing spaces through urban squatting emerged as a response to the violence of house evictions, the spreading of gentrification, anti-migration policies, and the collapse of the welfare state, one example being the squatting of abandoned hotels in downtown Athens to provide shelter for refugees. Through these occupations, new collective forms of everyday life flourish responding to the hardships of creative destruction and accumulation by dispossession, while serving as emancipatory sites that challenge the unyielding predetermination of lives and livelihoods as well as giving birth to the 'autonomous city' (Vasudevan 2015).

In this context, I define the social reproduction of resistance in the anti-authoritarian/anarchist commons in Athens as a mix of affective and connective labour practices that address the emotional needs and manual tasks necessary for the everyday context of collective actions and co-habitation/co-existence within community spaces. These practices include not only their logistics of maintenance and functioning, i.e. manual labour, but also the material non-manual labour that is vital for the enabling and maintenance of resistance, the formation of a sense of community, solidarity, inclusivity, and the production and maintenance of safe egalitarian spaces. The material manual labour that is embedded within social reproduction practices takes the form of collective cooking, cleaning, and the modification of occupied spaces and buildings, including the collecting and adapting of furniture, the collection of clothing and food for marginalized groups (such as the homeless or refugees), the printing and distribution of political material across the city, the patrolling of community events under the threat of police or neo-Nazi raids, graffiti writing, and the creation and running of communal libraries. The material non-manual labour takes the form of organizing events, the creation and maintenance of legal support and advice including fundraising for trial expenses, producing and transmitting knowledge (in the form of discussions, guest-lectures, and panels from academics-activists and/or members of foreign anarchist collectives), the creation and operation of information networks (in the form of anti-authoritarian websites, twitter accounts, self-defence classes), and providing healthcare in emergency situations (especially in the context of police brutality and riot violence).

The above labour practices tend to be coded through a gendered understanding of space and the gender roles performed within it, either feminine or masculine, often reproducing the gendered division of labour and space commonly seen in the Greek patriarchal household. The feminized labour of managing the 'household', in this case the everyday tasks of maintaining the physical spaces of the commons, are attributed to women as a more suitable (and possible) contribution to the struggle, while the heavy manual labour that is often necessary for the formation of the physical spaces of the commons is usually coded as masculine, along with direct action as the dominant form of protest. This division of labour and coding of resistance roles extends to the reproduction of a separate spheres model, one masculine and one feminine, with public space often seen as the space of heroic anti-authoritarian masculinity and direct action, in contrast to the sheltered physical (and online) spaces where women can seek refuge and perform a less 'direct', safer role of organizing or documenting.

As stated by K., a female anti-authoritarian, on the attempts to dismantle these gendered divides of labour within these collective spaces:

> Issues of who is doing what are always brought up in organizing meetings but that doesn't mean that it actually works. We make up some kind of rules about it but, for example, at the end of a party or a show you see especially men trying to avoid their assigned tasks. They either have a 'serious discussion' or are doing something else 'more important' leaving everything in a mess. At some point someone has to do it. You can't leave everything like that. Some of us actually live in those spaces. But every time there is some kind of excuse and, personally, it makes me feel bad that sometimes I am the one to yell about it and be the bad one.

As narrated by M., a non-binary member of another collective, sometimes the distinct roles are still considered to be determined by a natural inclination to a gender specific ability:

> Some years ago... in my previous collective... you could see women that considered it is their own duty to clean up, eying each other 'come on, we can do it quicker we know how to do it' because men in the group wouldn't do it... and it might seem funny now to some but we made a big effort for all those things to be written down as rules in operating the space.

The effort that M. stresses in their account encompasses the emotional labour that is performed by women and queer folk in order to create egalitarian spaces within the community, efforts that, as was repeated in interviews, resulted in a state of emotional and sometimes even physical 'exhaustion'. Often under the assumption that feminized tasks are not that difficult or 'dangerous', women and feminized subjects are under pressure to 'do more' in order to 'equally' contribute to the anti-authoritarian struggle.

The gendered division of tasks remains stubbornly de-politicized, understood as merely an issue of practical organization and not a political one. The emotional and physical impacts on female and queer members of struggling in a culture of resistance that is distinctly masculinist and heteronormative is also depoliticized. The gendered division of space that is reproduced by the gendered division of tasks reveals a masculinist coding of direct action and implies that socially reproductive feminized labour is at the bottom of the hierarchy of resistance practices. This results in the exclusion of women and feminized subjects from the struggle of reclaiming public space in the city as well as undermining the voicing of their demands in their own desired mode of resistance as less militant. The gender oppression that women and queer folk face in the anti-authoritarian community is thus rooted in the deeply patriarchal heteronormative structuring of Greek society and extends to the 'affective economies of hostility' within which anti-patriarchal, anti-authoritarian, and anti-neoliberal struggle has to be reproduced.

The Commons and the De-politicization of the Personal

The importance of the affective labour of women and queer folks in the anti-authoritarian community is made evident through the confronting of oppression, sexism, and gendered inequalities. The affective practice of claiming space-time and respect contributes to the remaking of (supposedly) safe egalitarian spaces and to the creation of support networks that defy and challenge the gendered economies of hostility that permeate the activist community.

On the 5th of October 2016, following an episode of sexual harassment of an immigrant woman by a Greek male member of the autonomous space Steki Metanaston, an autonomous group of lesbian activists, Lesviaki Omada Athinas (LOA),[16] left the social centre. Central to LOA's decision was not only the episode but also the sexist, homophobic, and patriarchal views that permeated Steki Metanaston and the reproduction of a masculinist attitude during the meeting in which the sexual harassment was to be discussed. In their announcement following the episode and the meeting, which was circulated around anti-authoritarian media and personal Facebook accounts, LOA stated:

> The words of this woman, our words and the words of all those who supported her were faced with irony and a constant humiliation. The assembly, through the course of our presence there, confirmed the straight male privilege that dominated the space without leaving any space for the expression, the lived experience and the effort to open up a dialogue on the issue of sexism towards the woman, towards the immigrant...

In this account, Steki Metanaston emerges as an unsafe, masculinist space where gendered hierarchies contribute to lived experiences of oppression, in turn reproducing exclusion and power relations that disrupt any meaning of solidarity and inclusivity. The above episode came as no surprise to the community, not least because, in the context of socializing within the space, violations of personal boundaries, sexist behaviours –

mainly in the form of micro-aggressions – often overlapped with an assumption of consent originating in rape culture and the performance of militant masculinities.

During one of the interviews, such patterns of harassment were brought up by S., a female anti-authoritarian, as a set of behaviours that were often seen within the spaces of collectives resulting in the formation of a hostile environment for female and queer anti-authoritarians:

> How many times have I seen this? It's always a one-time-incident. It won't happen again, he is a good guy, he was a bit drunk and everyone was having fun. And then you end up being either the slutty one or the super-sensitive feminist.

The frequent occurrence of such 'one-time-incidents', however, had cemented feelings of violation of personal boundaries and of the safety of collective spaces, as well as having contributed to an overall feeling of disrespect towards women and queer folks. The impact of such sexist and masculinist attitudes that violate consent have also resulted in rapes that are rarely openly acknowledged as political issues and forms of hostility that engender fear and ultimately lead to violent forms of exclusion.

Female and queer anti-authoritarians find themselves situated within a spatial economy of violence and hostility that impacts their mobility and produces spaces of resistance in which they face multiple, intersecting, and also contradictory sources of violence and oppression. Within this spatial economy, ethical and moral dilemmas are generated in relation to political positioning and identities (across, gender, race, class, and sexuality). For example, the stigmatization of harassment victims is often equated with a moral contradiction, i.e. that of putting the individual before the collective, supposedly distracting activists from the greater cause of struggle against all forms of power and disturbing the unity of groups and collectives.

As N., a young female anarchist, explained to me in an interview, the naming of sexual harassment and abuse is itself a complex process. Noting that previous experiences of such incidents within the community have resulted in her being discouraged from speaking up about these incidents, N. also linked them to conceptualizations of the 'good' and 'bad' female anarchists, the former being one who can confront men on a personal level but at the same time be 'faithful' and 'understanding' towards the anti-authoritarian community. In this context of the 'femi-nazi' and 'political-correctness' rhetoric, the labels of feminism and anti-sexism are often avoided by some, although proudly reclaimed by others. N. explained that being young (20-years old) and relatively 'new' to the community, she did not know how to confront the men who were touching her body without her permission while being surrounded by people that she would often meet in political actions and events. She felt 'puzzled' by the fact that the act of 'calling out' or confronting someone led to her being labelled as either a 'good' or 'bad' anti-authoritarian woman:

> Sometimes it's impossible, I don't know what to do. There are even women in the community that write whole texts against 'anti-sexism' that makes us victims and that if we are basically true anarchists we will fight back, we won't be a victim that cries or is upset. And how many times can you do that? It's not like that.

In this spatial economy of sexism and hostility, it is difficult for female and queer anti-authoritarians to name oppression in the context of a community in which male privilege underlies every collective process and understandings of struggle.

In the spaces of resistance and activism against authority and police brutality, sexist, homophobic, and abusive behaviours and rhetoric abound, emulating the very capitalist urban spaces they critique. These behaviours are not criticized but rather ignored or dismissed as 'personal conflicts or individual obsessions' and 'issues' that might 'harm' the community (and its image). As a result, the experiences and trauma inflicted by these behaviours ostracizes groups and individuals, depoliticizing and normalizing gender-based violence in the community. Thus, sexist and homophobic behaviours, sexual harassment, and intimate partner violence performed by male anti-authoritarians is either encouraged to be 'kept in private' as a personal issue or is subjected to a comparative 'evaluation' of the abuser's and the victim's contribution to the movement, serving to claim that the anti-fascistic and anti-authoritarian struggle is more important and political than the anti-patriarchal one.

N., a male anti-authoritarian, discussing the reaction around intimate partner violence within the community elaborates on the supposed personal character of such issues:

> How many issues can this meeting be dealing with, honestly? Whatever people do in their relationships is their own business, and there is no reason for us to apologize for the actions of just one man because we are men too. People get drunk and have issues and get angry and don't know how to control themselves ... but you know she is a grown-up woman too, she can fight back ... of course everyone is against these kind of behaviours but we are not going to end up being cops in anyone's life.

On the other hand, the account of L., a female activist who was the victim of violence by her partner, who was also active in the anti-authoritarian struggle, illustrates the cost of such behaviour in the context of everyday life in the spaces of the community:

> I tried to completely cut him out of my life but I couldn't just not attend events or meetings. I was seeing him around sometimes and it caused me great anxiety to the point that I had to openly talk about it with people in the community that we both were friends with ... some of them thought it was very difficult to believe ... others said I should keep this to myself. Eventually I stopped attending actions that I knew he was going to attend too. But my life was built around all those that it was very difficult to walk away. I started hanging out with people outside of the community too.

In certain cases highlighted in the interviews, when such episodes were made public in the community, there was a gradual collapse of support networks around survivors, which led to their exile from the community in order to deal with and, hopefully, heal from the traumatic experience. Other times, abuse survivors found solidarity amongst feminist and queer anti-authoritarians who were willing to support them during the process of openly talking about the abuse, which, in turn, formed networks of care and protection (as in LOA's case).

While anti-authoritarian action and practices are met with violent state, police, and far-right responses in the form of raids, attacks, and brutality, the normalized hostile masculinist environment within the anti-authoritarian community generates even more distress to women and queer folks who are fighting a struggle in multiple directions, exposed to precariousness, vulnerability, and dispossession. The co-existence of survivors and allies with abusers within the spaces of the community, combined with the symbolic erasure of their traumatic experience, greatly impacted upon their experience of space, forming geographies of fear, anger, and isolation in supposedly safe spaces. Emotional survival within a masculinist society and the effort to reveal and disrupt patterns of sexual harassment and gendered violence, both in the activist commons and beyond in the city, are enacted through the formation of support networks and practices by survivors and allies. These networks and practices provide comfort, trust, and courage, which mutually enables resistance and allows for a reclaiming of space within the geographies of militant masculine performance on the one hand and the patriarchal capitalist violence of the neoliberalizing city on the other.

Anarchist Commons: Performances and Cultures of Resistance and the Re-making of Safe Spaces

One issue that kept on surfacing in the interviews, and which is also reflected in feminist and queer anti-authoritarian political texts, is the refusal to acknowledge the hostility that sexist, homophobic, and hegemonic masculinist behaviours, produce in social spaces. Spaces of resistance are made 'safe' only for males and those who gain respectability and trust in the community by their proximity to certain subjectivities, namely what Coleman and Bassi (2011, pp. 212–216) refer to as the 'Anarchist Action Man' and 'Man with Analysis', in order to describe the hyper masculine and violent performativity of direct action (Sullivan 2005). This dynamic reinforces the authority and status of older male members within the community. These masculinities, connected to activist cultures that set the almost impossible (gendered) standard of the 'ideal activist' (Craddock 2019, p. 138) are reproduced in the commons often resulting in 'spaces of resistance reinforcing dominant oppressive structures' (Craddock 2019, p. 140), generating feelings of uncertainty, and isolation along with an atmosphere of hostility.

In the context of the anti-authoritarian movement in Greece, the hegemonic figure that emerges is that of the 'anarcho-father' (Boukalas 2011; Apoifis 2017), a man in his 50s to 60s who has been active in the struggle for decades and who thus has the privilege to 'patronize, explain to you theories that you are supposed to not get, [and] make sexist jokes because things were different in the old days' as observed by N. while describing the authoritative power and status that such 'anarcho-fathers' enjoy within the community. The figure of the anarcho-father, thus, embodies not only the masculinist patriarchal structuring of Greek society but also the extent of gender oppression in the city, which has permeated supposedly safe egalitarian spaces. The spatialities of gender roles of a heteronormative familial society that centres the authoritative father figure are reproduced, normalizing silencing and abuse of power on the basis of gender and sexuality. This gendered structuring dictates the limits of political participation, the spatio-temporality of visibility of women and queers in the city, and one's 'proper place' in the physical and social spaces of the urban.

Violent, aggressive, sexist, and hegemonic behaviours and masculinist stylization of the body and speech regularly appear in the anti-authoritarian community as the 'correct' or 'proper' way to perform political and cultural identity. In turn, these political subjects gain social power, normative authority, and prestige in the activist community, placing them at the top of unofficial hierarchies. In the spaces of everyday life in the community, these masculinities are performed through micro-aggressions in communication, especially in the course of collective organizing meetings and assemblies, through adopting an aggressive tone and posture, raising of the voice while another individual with less experience or specific gender identification (especially women and queer folks) is talking, demonstration of knowledge and expertise around resistance practices, political theory, and infrastructure building/organizing areas of knowledge that are coded as masculine, as well as the extensive use of gendered and sexually charged language when referring to the state, the police, employers, and neo-Nazis. The gendered language of political protest is not only evident in the speech of the Anarchist Action Men but also in the visual protest of graffiti writing across the city (and especially in the heavily policed downtown area), stickers, posters, and flyers, during protests as part of the chants, during arguments in the assemblies, in the lyrics of the songs that frame the resistance, and in the everyday interpersonal political discussions in the everyday life in the community. The 'cops' are 'faggots', the rape of 'golden dawn women' is seen as a form of anti-racist justice, the state and employers are called 'whores' and 'pimps', fear is equated with being a woman, and demands for freedom are mixed with sex and violence as gendered 'weapons' of resistance.

The issue of the violent sexist language used in political anti-authoritarian contexts was frequently brought up in the interviews as part of the spatialized affective economies of hostility, generating feelings of anxiety, frustration, disrespect, and (un)belonging. In the statement below, following a performance filled with heavily sexualized and violent language during the anti-authoritarian and queer organized 3rd Sexuality Festival, the issues of (un)safe spaces, representation, and inclusivity were brought up in a statement by feminist and queer anti-authoritarians:

> We were thinking about the individuals we wish to give space to in our shows and our spaces. And even more, we were thinking about the individuals we exclude from these spaces and under what circumstances that happens. We were wondering if a woman who has been harassed or abused, if an immigrant who has been bashed in the street could stay throughout the show and listen to all this. Above all, we were wondering why someone like them would have to experience all this within our spaces.

Explaining that what counts as shocking and revolutionary – 'just words and rhymes' for the white, straight, able-bodied Greek anti-authoritarian figure – might be for others in the same spaces, a lived traumatic experience of exclusion, oppression, and violence. These 'others', seeking refuge from the already hostile neoliberal environment of the city, are not only looking for safety but also a community and support (emotional or material). Confronting individuals who articulated sexist, homophobic behaviours within the commons, demanding and constantly highlighting the importance of a gender-neutral allocation of everyday tasks and labour, as well as the inclusion of gender-related agendas, are some of the practices that ensure safety and egalitarianism within the anti-authoritarian commons. The support, the sense of community, and safety are becoming possible through affective labour performed in the anti-authoritarian commons that ensures and demands a genuinely safe and transformative place against both homophobic masculinist behaviours and revanchist neoliberal governance.

Politicizing Emotion: Dispossession and Empowering Practices of Social Reproduction in the Urban

Perceived often as 'failed' subjects within cultures of aggressive, hyper-masculine, and direct-action oriented political engagement and activism, women, queers, and feminized subjects perform a 'queer art of failure' (echoing Halberstam 2013),[17] creating ethics of care and community by failing to measure up to and by challenging patriarchal ideals of evaluating success and contribution to the struggle. Marginal both in the context of the neoliberal city and their political communities, the 'failed' subjects may occupy the lower ranks of socio-political and performative hierarchies, but their choice to resist from that complex marginal position emerges as deeply political queer feminist praxis. This intentional 'failure' reveals threads of comradeship and material and emotional acts of care that transcend the boundaries of hegemonic political performativity and fosters a queer feminist anti-authoritarian resistance on multiple levels. These threads form the building blocks for a struggle against dispossession that refuses the allocation of political subjects into 'one's proper place': through their unruly presence in events where they are commonly perceived as (in)visible; through the 'calling out' of abusive and oppressive behaviours; and from the possibilities of a future that can be truly and not partially revolutionary.

In the social reproduction of resistance, what emerges as vital is the emotional labour of collective and self-care between those who reclaim space in the context of multiple contradictory margins, and their resilience and persistence in 'carving up' safe spaces that form pedagogies of solidarity, modes of survival, and reproduce resistance by transforming struggle. This collective and self-care mainly informed through informal networks between women and queers has included discussions about mental health issues, the choice of treatment of which often becomes a political and ethical issue within the community. Within a context of an activist culture of 'martyrdom' (Chen and Gorski 2015, p. 379), self-care is framed as an individualistic neoliberal personal strategy to serve capitalistic productivity (disconnecting it completely from the wider political context in which it occurs) and is often expressed as a rejection of psychiatric medication and assistance. As a result, the practice of the management of stress and panic disorders in stressful situations are being dealt with through the queer feminist practice of 'having your back' and 'checking in on each other' against the isolation of such feelings as 'personal'.

The spatiality of these practices has ranged from the reclaiming of social and physical spaces, such as 'never again walking alone home at night', to forming small informal groups (as 'friends') to attend events (including protests) that otherwise seemed intimidating and stressful because of the hyper masculine behaviours of anti-authoritarian men on the one hand and the fear of police brutality on the other. The above practices also signify an effort to produce spaces in which non-normatively and feminized gendered subjectivities can be expressed in empowering ways, in turn enabling more effective forms of community and resistance to the neoliberalization of the city.

As witnessed both during the interviews and observation throughout my own personal experiences in anti-authoritarian spaces, not all of the above practices have been successful in terms of shifting the masculinism of anti-authoritarian resistance, as I., a queer anti-authoritarian, noted:

> The truth is that we have to be persistent and it feels like we are having the same discussions again and again sometimes, but sexism and homophobia and the privileges of being a man cannot be unlearned in just some meeting or because someone read Judith Butler [laughing]...

Other interviewees, such as N., a female anti-authoritarian, expressed their concern that in this process of 'unlearning' male privilege and including gender-related agendas, women and queers end up being the 'representatives' of gender justice without any genuine engagement coming from the part of other members, who are profiting from the creation of a more progressive and inclusive profile within the community:

> It's not only that whenever issues related to gender come up some show indifference and leave it to us ... it's also that some have this attitude of 'educate me' as if it is our only job to teach them how they should be treating someone with respect ... as if they don't know how to do that.

Despite the difficulties faced, it is through the practices of document-ing, archiving, and making known issues of sexism, homophobia, and abuse within the anti-authoritarian community, that feminists, women, and queer anti-authoritarians produce knowledge that forces their com-munity as a whole to face the oppression that it reproduces. Beyond rais-ing awareness, these collective self-care practices also form memory and shared ties that enable the transformation of repertoires of resistance themselves. Insisting on bringing up issues of oppression, even if this means a loss of their support network or of one's sense of identity and community, is a deeply political act. In the context of queer and feminist collective self-care, naming a hidden oppression empowers one to be not only privately but publicly vulnerable in a social and physical space where hegemonic performativity too often dictates the boundaries and conditions of 'the political'. Additionally, such practices require re-im-agining and re-forming networks of healing, solidarity, and well-being.

Through the efforts of confronting and documenting gender oppres-sion, the masculinist narrative of harassment and sexism as isolated inci-dents is being subverted, ultimately resulting in the disruption of power hierarchies, their naming and recognition as such, and the re-claiming of the spatialities of social reproduction and the body as political. I argue that the processes of disobedience and opposition to masculinist activ-ism are empowering ones, as they 'force' a collective renegotiation that can enable the formation of new collectivities, and the re-articulation of claims and practices, and a new approach to the social reproduction of resistance.

The embodied resistance of 'failed' political subjects emerges from within obscured geographies of oppression through the acts of reclaiming and re-producing space. Ultimately, the social and spatial reproduction of resistance disrupts and subverts the symbolic violence and erasure caused by the non-recognition of oppression and its invisibility, which in turn threatens the sense of agency and credibility of survivors.

The practices of social reproduction that are documented here are not spectacular in a sense that they do not refer to a singular violent clash which dominates the collective imaginary, such as that of the occasional street-war against the police, but they make resistance possible through the constant struggle of emotional survival, emotional labour, care, sup-port, and re-negotiation of the idea of community that enables an alternate future. This alternate future also socially reproduces space and commu-nity for those 'others' that seem not to belong in the neoliberal vision of a creative, tourist city and whose life is deemed unliveable and unworthy by the necropolitics of patriarchal capitalism that condemn the non-white, non-able bodied, non-heteronormative, and non-male disposed subject.

Conclusion

In this chapter, I have showcased, through a case study of everyday life in the anti-authoritarian/anarchist spaces of Athens, that affective labour, visibility, practices of collective (and self-)care, and well-being within activist communities are gendered issues, vital to the social reproduction of resistance, in the context of intensifying violent, sexist, racist, and classist neoliberal urban governance. Anti-patriarchal struggle is not to be dealt with 'after the revolution', thereby creating hierarchical experiences of oppression and axes of difference in an economy of struggle.

The practices of care and kinship developed by 'failed' political subjects are mechanisms of emotional survival and well-being that, within their community, reveal the significance of social reproduction as a political praxis. Without this praxis, no revolutionary call for action can be achieved. Thus, these practices do not function as destructive for the movement but as politically radical in that they indicate the possibility of alternate futures, through the re-articulation and re-formation of the ground upon which oppression is built. They constitute the coming into being and living life outside and against social hierarchies, economies of struggle, and geographies of domination that characterize the neoliberal city 'in crisis'.

The multiple axes of resistance embedded within the socially reproductive feminized practices of urban anarchist commoning, embodying a feminist ethics of care, solidarity, and mutual aid, indicate the importance of gender at the heart of resistance and of the political. Racism, gendered violence, sexism, homophobia, and misogyny intensify affective economies of hostility and the diffusion of such hostility within spaces of resistance only highlights the necessity of the (re)production and reclaiming of space. An intersectional feminist praxis enables resistance not only by creating spaces of radical resilience that are safe and truly egalitarian, not only within the neoliberal city, but also in the commons. The material and affective, emotional, and connective practices of social reproduction thus inform the visibility and pedagogy of gendered struggle for a feminist and queer right to the city.

Acknowledgements

I would like to thank the editors for their constructive guidance and critique throughout the writing of this chapter.

Notes

1 See Boukalas 2011; Vradis and Dalakoglou 2011; Apoifis 2017; Cappuccini 2017; Daskalaki 2018.

2 There are, though, notable exceptions (see Marinoudi 2017, 2018).

3 The murder of student Alexis Grigoropoulos on December 2008 in the central neighbourhood of Exarchia fuelled a wave of youth protest and riots across the country.

4 Participants requested not to use their real names or any other defining characteristics.

5 See Dalakoglou 2013; Carastathis 2015; Eleftheriadis 2015; Kotouza 2015; Athanasiou and Kolocotroni 2018.

6 The economic crisis of the downtown commercial district, which was accelerated during the crisis due to state policies, extensive taxation and a competitive real estate market was framed in political discourses as caused by extensive protests and the migrant populations residing in the already low-quality buildings in the area. The discourse of the 'ghettoization' of the downtown was intertwined with discourses of racialized crime and sex work as well as the criminalization of protest.

7 'Neighbourhood patrols' formed by members of the far-right neo-Nazi party 'Golden Dawn' extensively harassed immigrants living in downtown areas such as Kypseli, including beatings, street bashing, destruction of property and business, as well as the lockdown of local playgrounds to migrant children.

8 Golden Dawn is a far-right neo-Nazi party.

9 At the same time, in Athens, an 'apolitical' version of DIY urbanism, often branded as 'participatory governance', has been celebrated by the state in the form of the encouragement of small scale 'creative' interventions in the city that aim to improve neighbourhood functions and the aesthetics of the urban environment. 'The right to the city' is increasingly perceived as the right of a young, able-bodied, creative class to be in a 'safe' city free of protests, refugees, immigrants, and 'annoying' leftists and anarchists (see Makrygianni and Tsavdaroglou 2015), built on the dominant heteronormative ideas for an ideal citizen-consumer, beneficial to the struggling national economy.

10 See Kambouri and Zavos 2010; Athanasiou 2014; Markantonatou 2015.

11 See Athanasiou 2014; Carastathis 2015; Makrygianni and Tsavdaroglou 2015; Markantonatou 2015.

12 'Greek boy' or *palikari (παλικάρι)*, an ironic figure of speech, can be characterized as a misogynistic embodiment of Greek heteronormative masculinity linked with the reproduction of homophobia, sexism, rape culture, slut shamming, racism, and white Greek masculinist supremacy.

13 Within a framework of anarchist/anti-authoritarian commoning emerges a political culture of resistance based on 'solidarity and forefront organizing through confrontation and construction; value-practices in affinity group organizing; and anti-oppression consciousness integrated into day-to-day

organizing' (Jeppesen et al. 2014, p. 884).

14   See Hoschchild 1979, 2012, 2013; Hardt and Negri 2004; Weeks 2007.

15   See Bhattacharjya et al. 2013; Kennelly 2014; Chen and Gorski 2015; Craddock 2019.

16   Literally called the migrants' house, Steki Metanaston is a place where various anti-authoritarian collectives are based including anti-racist, anti-fascist, and leftist initiatives, as well as refugee, migrants and LGBTQ groups.

17   As Halberstam (2013, p. 4) stresses 'Where feminine success is always measured by male standards, and gender failure often means being relieved of the pressure to measure up to patriarchal ideal, not succeeding at womanhood can offer unexpected pleasures'.

References

Apoifis, N. (2017). *Anarchy in Athens: An Ethnography of Militancy, Emotions and Violence*. Oxford: Oxford University Press.

Athanasiou, A. (2014). Precarious intensities: Gendered bodies in the streets and squares of Greece. *Signs: Journal of Women in Culture and Society* 40 (1): 1–9.

Athanasiou, A. and Kolocotroni, V. (2018). On the politics of queer resistance and survival: Athena Athanasiou in conversation with Vassiliki Kolocotroni and Dimitris Papanikolaou. *Journal of Greek Media and Culture* 4 (2): 269–280.

Bhattacharjya, M., Birchall, J., Caro, P. et al. (2013). Why gender matters in activism: Feminism and social justice movements. *Gender & Development* 21 (2): 277–293.

Boler, M., Macdonald, A., Nitsou, C. et al. (2014). Connective labor and social media: Women's roles in the 'leaderless' Occupy movement. *Convergence* 20 (4): 438–460.

Boukalas, C. (2011). No one is revolutionary until the revolution! A long, hard reflection on Athenian anarchy through the prism of a Burning Bank. In: *Revolt and Crisis in Greece: Between a Present yet to Pass and a Future Still to Come* (eds. A. Vradis and D. Dalakoglou), 279–298. Chico, CA: AK Press.

Cappuccini, M. (2017). *Austerity & Democracy in Athens: Crisis and Community in Exarchia*. Cham, CH: Springer International Publishing.

Carastathis, A. (2015). The politics of austerity and the affective economy of hostility: Racialised gendered violence and crises of belonging in Greece. *Feminist Review* 109 (1): 73–95.

Chen, C.W. and Gorski, P.C. (2015). Burnout in social justice and human rights activists: Symptoms, causes and implications. *Journal of Human Rights Practice* 7 (3): 366–390.

Coleman, L.M. and Bassi, S.A. (2011). Deconstructing militant manhood: Masculinities in the disciplining of (anti-)globalization politics. *International Feminist Journal of Politics* 13 (2): 204–224. doi:10.1080/1 4616742.2011.560039

Craddock, E. (2019). Doing 'enough' of the 'right' thing: The gendered dimension of the 'ideal activist' identity and its negative emotional consequences. *Social Movement Studies* 18 (2): 137–153.

Dalakoglou, D. (2013). The crisis before 'the crisis': Violence and urban neoliberalization in Athens. *Social Justice* 39 (1): 24–42.

Daskalaki, M. (2018). Alternative organizing in times of crisis: Resistance assemblages and socio-spatial solidarity. *European Urban and Regional Studies* 25 (2): 155–170.

Eleftheriadis, K. (2015). Queer responses to austerity: Insights from the Greece of crisis. *ACME: An International E-Journal for Critical Geographies* 14 (4): 1032–1057.

Halberstam, J. (2013). *The Queer Art of Failure*. Durham, NC: Duke University Press.

Hardt, M. and Negri, A. (2004). *Multitude: War and Democracy in the Age of Empire*. London: Penguin.

Hochschild, A.R. (1979). Emotion work, feeling rules, and social structure. *American Journal of Sociology*, 85 (3): 551–575. http://www.jstor.org/stable/2778583

Hochschild, A.R. (2012). *The Managed Heart: Commercialization of Human Feeling*. Updated edition. Berkeley: University of California Press.

Hochschild, A.R. (2013). *The Outsourced Self: What Happens When We Pay Others to Live Our Lives for Us*. London: Picador.

Jeppesen, S., Kruzynski, A., Sarrasin, R. et al. (2014). The anarchist commons. *Ephemera* 14 (4): 879–900.

Johansson, A. and Vinthagen, S. (2016). Dimensions of everyday resistance: An analytical framework. *Critical Sociology*, 42 (3): 417–435.

Kambouri, N. and Zavos, A. (2010). On the frontiers of citizenship: Considering the case of Konstantina Kuneva and the intersections between gender, migration and labour in Greece. *Feminist Review* 94 (1): 148–155.

Kennelly, J. (2014). 'It's this pain in my heart that won't let me stop': Gendered affect, webs of relations, and young women's activism. *Feminist Theory* 15 (3): 241–260.

Kotouza, D. (2015). *Surplus Citizens: Struggles in the Greek Crisis, 2010–2014* (Doctoral dissertation, University of Kent).

Kousis, M. (2017). Alternative forms of resilience confronting Hard Economic Times. A South European perspective. *Partecipazione E Conflitto* 10 (1): 119–135.

Makrygianni, V. and Tsavdaroglou, C. (2015). The right to the city. Athens during a crisis era: Between inversion, assimilation and going beyond. In: *City of Crisis: The Multiple Contestation of Southern European Cities* (eds. F. Eckardt and J.R. Sànchez), 29–57. Bielefeld, DE: transcript Verlag.

Marinoudi, S. (2017). *I zoi choris emena: Emfyla ypokeimena entos ki ektos ton kinimatikon choron*. Athens: Futura.

Marinoudi, S. (2018). Queer subjectivities within political scenes: Traumatic relations, exposed vulnerabilities. *Journal of Greek Media & Culture* 4 (2): 151–166.

Markantonatou, M. (2015). State repression, social resistance and the politicization of public space in Greece under fiscal adjustment. In: *City of Crisis: The Multiple Contestation of Southern European Cities* (eds. F. Eckardt and J.R. Sànchez), 199–212. Bielefeld, DE: transcript Verlag.

Mott, C. (2018). Building relationships within difference: An anarcha-feminist approach to the micropolitics of solidarity. *Annals of the American Association of Geographers* 108 (2): 424–433.

Oksala, J. (2016). Affective labor and feminist politics. *Signs: Journal of Women in Culture and Society* 41 (2): 281–303. doi:10.1086/682920

Sears, A. (2017). Body politics: The social reproduction of sexualities. In: *Social Reproduction Theory: Remapping Class, Recentering Oppression* (ed. T. Bhattacharya), 171–191. London: Pluto Press.

Sullivan, S. (2005, February). *'Viva nihilism!' On militancy and machismo in (anti-)globalisation protest.* (Centre for the Study of Globalisation and Regionalisation Working papers No. 158/05). http://www2.warwick.ac.uk/fac/soc/csgr/research/workingpapers/2005/wp15805.pdf

Tsavdaroglou, C. (2016), Urban commons and the right to ambiance: Gentrification Policies And Urban Social Movements in Barcelona, Athens and Istanbul. *Ambiances, tomorrow. Proceedings of 3rd International Congress on Ambiances. September 2016, Volos, Greece*, Volos, Greece. pp. 707–712. ⟨hal–01414154⟩.

Vaiou, D. (2014). Is the crisis in Athens (also) gendered?: Facets of access and (in)visibility in everyday public spaces. *City* 18 (45): 533–537.

Vaiou, D. (2016). Tracing aspects of the Greek crisis in Athens: Putting women in the picture. *European Urban and Regional Studies* 23 (3): 220–230.

Vaiou, D. and Kalandides, A. (2017). Practices of solidarity in Athens: Reconfigurations of public space and urban citizenship. *Citizenship Studies* 21 (4): 440–454.

Vasudevan, A. (2015). The autonomous city: Towards a critical geography of occupation. *Progress in Human Geography* 39 (3): 316–337.

Vradis, A. and Dalakoglou, D. eds. (2011). *Revolt and Crisis in Greece: Between a Present yet to Pass and a Future Still to Come.* Chico, CA: AK Press.

Weeks, K. (2007). Life within and against work: Affective labor, feminist critique, and post-Fordist politics. *Ephemera: Theory and Politics in Organization* 7 (1): 233–249.

5

'Sustaining Lives is What Matters'

Contested Infrastructure, Social Reproduction, and Feminist Urban Praxis in Catalonia

James Angel (King's College London)

Introduction

Esther described herself to me as 'bolshie for as long as I can remember, in every way.' A single mother from El Vendrell, a touristy town in the Catalan province of Tarragona (south-west along the coast from Barcelona), Esther is one of millions within Spain whose daily reproductive practices have become highly precarious in the wake of the crisis of financialized capitalism that has rocked the country since 2008 (Charnock, Purcell, and Ribera-Fumaz 2014). With unemployment endemic and austerity measures biting, Esther, like many others in Catalonia, is forced to rely upon illegally occupied housing and tampered access to electricity, gas, and water networks to get by. In this sense, the reconfiguration of the urban environment via processes such as state austerity alters human-infrastructure relations, manifesting in a crisis of social reproduction. Yet the argument in this chapter is that just as urban change shifts reproductive practices, equally, by reproducing ourselves differently via adjusting human-infrastructure relations, we can remake the city. The chapter draws on four months of ethnographic research[1] conducted in Catalonia between 2016 and 2018, within an activist network called la

A Feminist Urban Theory for our Time: Rethinking Social Reproduction and the Urban, First Edition. Edited by Linda Peake, Elsa Koleth, Gökbörü Sarp Tanyildiz, Rajyashree N. Reddy & darren patrick/dp.

Alianza Contra la Pobreza Energética (translated as the Alliance Against Energy Poverty, hereafter referred to as APE, pronounced 'ap-eh').[2]

My intention, in accordance with the broader project of this book, is first to contribute to the important task of deepening conversations between urban studies and materialist-feminist theorizations of social reproduction. Yet I also perceive and seek to respond to a further under-explored question within emerging geographical engagements with social reproduction, which is the potential of social reproduction as a starting point for a feminist praxis capable of realizing socially just and liberatory futures. The feminist usage of the concept of social reproduction, of course, emerged through political struggle, within the Wages for Housework campaign of the 1970s (Dalla Costa and James 1975). By naming as 'work' the racialized and gendered processes required to produce labour power as a commodity, feminist activists sought to make a major intervention within dominant conceptions of the strategies, subjects, and demands of emancipatory politics (Weeks 2011). Recent geographical engagements with social reproduction have retained an understanding of reproductive practices as a terrain of struggle in which relations of violence, domination, and exploitation might be perpetuated and/or resisted (Mitchell, Marston, and Katz 2004; Meehan and Strauss 2015). Yet the question foregrounded by Silvia Federici (2012) of how we might reproduce ourselves differently and, in turn, begin to craft alternative socio-ecological trajectories, has perhaps not been at the forefront of contemporary geographical debates. The trajectories of the commons imagined by Federici (2012) are not necessarily specifically urban. I hope to illustrate, though, that Federici's provocation towards an understanding of social reproduction as a basis for emancipatory struggle is generative for building feminist urban theory and, in particular, a feminist urban theory that foregrounds a dialectical relationship to feminist urban praxis.

APE's struggles against urban infrastructure disconnections form the basis of this argument. Founded in Barcelona in early 2014, local groups of the APE network have since been formed in towns and cities across Catalonia. APE was established by activists who had, for some years, been working on campaigns that sought to challenge the privatization of water and energy in Barcelona. With the issue of water and energy disconnections becoming far more acute since the global financial crisis (Tirado Herrero and Jiménez Meneses 2016), and increasingly prominent in mainstream political discourse, the decision was taken to focus on this issue as a campaigning focus. Since its formation, influenced by the approach of anti-eviction movement la PAH (la Plataforma de Afectadas por la Hipoteca, the Platform for People Affected by Mortgages) (see Gonick 2016a; García-Lamarca 2017), APE has developed a form of political organizing premised upon collectively contesting disconnections from water, electricity, and gas networks, framing their campaign around the right to what they term the 'basic supplies' required for a dignified life.

APE, I will illustrate, enacts a feminist praxis premised upon the creation of more caring and collectivized modalities of social reproduction. The reconfiguration of social reproduction portended by APE, I will argue, is the departure point for the potential reconfiguration of the urban process more broadly. As such, this chapter contributes to ongoing attempts to develop a 'social ontology' of the urban according to which daily reproductive practices are not the mere effect of an all-encompassing capitalist urbanization process but, rather, are actively constitutive of the urban environment (Ruddick et al. 2017). Specifically, I attend to the role of urban infrastructure within this social ontology – with 'infrastructure' understood, following AbdouMaliq Simone's (2004b) claim that people can become infrastructure, as inclusive of an array of human and non-human practices and processes implicated in reproducing urban life.

This chapter begins by situating my own relationship to the debates and struggles in question, tracing some of the questions and frustration raised by my pursuit of a scholar-activist ethnographic method. I then move on to establish the conceptual foundations necessary for my argument to proceed, offering some preliminary thoughts on the relationship between social reproduction, the urban, and infrastructure. Finally, I proceed to contextualize the Catalan study, before turning to the infrastructural struggles of APE and their relevance for the question of social reproduction and the urban.

Positionality and Praxis

The argument within the chapter draws upon four months of ethnographic research conducted in Catalonia between 2016 and 2018. Three months of this research was undertaken in April, May, and June 2017, alongside two weeks of preliminary fieldwork in October 2016 and two weeks of follow-up research in February 2018. My research consisted, primarily, of participant observation in group meetings, events, and protests. This was complemented with analysis of relevant documents such as campaign blog posts, promotional materials, and media articles, alongside a series of 36 semi-structured interviews with activists (from APE and other related activist networks), politicians, civil servants, energy industry consultants, and representatives of private firms.

Feminist scholarship's insistent questions about identity, positionality, and reflexivity within the research process have helped unsettle masculinist and positivist ideas around 'objective' knowledge-production through a recognition of all knowledge-claims as always-already situated (Haraway

1988) and enmeshed within specific historical geographies (Katz 2001a). In this regard, in terms of my own relationship to the Catalan energy struggles under discussion, I have no pretence of impartiality. Rather, my ethnographic involvement with APE constitutes one component of a broader doctoral research project, which I am endeavouring to integrate into a scholar-activist praxis (Autonomous Geographies Collective 2010; Halvorsen 2015). My research emerges out of a decade of participation within various activist initiatives and social movements focused around climate justice (see Chatterton, Featherstone, and Routledge 2013) and, in more recent years, the specific question of renewable energy transition (see Bridge et al. 2013). Indeed, I first became interested in APE's work through my encounters with APE activists within international energy activist networks. After attending (in an activist capacity) a conference on solutions to energy poverty hosted by APE in Barcelona, October 2016, APE activists extended an invitation to me to return for a sustained period of research – an invitation that I was very happy to accept.

My approach to navigating my dual role as both 'activist' and 'scholar' is, in the terminology of Derickson and Routledge (2015), to attempt to 'resource' the struggles I work within, both materially and epistemologically. The aim here is to avoid extractive research practices that instrumentalize movements in order to further academic careers, and to refrain from patronizing attempts to offer 'enlightened' strategic guidance from on high. A scholar-activist praxis of resourcing, instead, begins with the question of what movements actually want, whether this be time, funding, meeting space, answers to particular research questions, or something else.

My capacity, as a scholar-activist, to resource APE was, however, restricted in a number of respects. First, my participation in the campaign took place over a time-bound period, and it took considerable time to build trust and relationships with my collaborators. Furthermore, I was operating in a political context that I was unfamiliar with and working outside of my first language. The result of these complexities was that it took my collaborators and I a while to establish the kinds of resources that I could feasibly and constructively provide. Ultimately (beyond my presence as an additional participant at demonstrations and leafleting sessions) this proved mainly to be through taking on written English translation work for the group – a role that I have maintained since my departure from Catalonia.

The silver lining here, however, is that my experiences within APE have helped fashion connections, solidarities, and learning between struggles. Richa Nagar (2002), in her reflections on transnational feminist praxis, argues that 'researchers must seriously consider how we can serve as useful channels of communication between scholars and activists located in different places' (p. 185). I learned a lot from my time within APE, including many strategic insights (for instance around the collective case-work approach discussed below) that I have shared amongst the UK activist networks I participate within on an ongoing basis. Moreover, I have helped to mobilize these UK networks in support of APE when fruitful, for example contacting various UK activist groups to arrange their support for an open letter of solidarity with APE following a legal case brought against several APE activists by water firm Aigües de Barcelona.

As well as the concern around extracting scholarly value from political struggle, I also had to grapple with the risk of instrumentalizing the lives of individual APE participants, many of whom (as will be described in greater depth) are living highly precariously in the face of violent austerity measures and a highly unjust urban infrastructural politics. Yet, at the heart of APE's praxis is a commitment towards combating the shame, atomization, and loneliness of poverty and neoliberal urbanism – which often translates as encouraging people to share their stories of personal struggle and loss with others. Accordingly, as I discussed my project more with APE participants, it became clear that many were eager to share their experiences with me and to see these represented in scholarly texts.

I will turn to discuss these experiences in more depth soon. Before doing so, however, I wish to clarify the conceptual implications of the struggles in question.

Social Reproduction, Infrastructure, and the Urban

As introduced already, the contribution I seek to make in this chapter is to excavate the relationship between infrastructure, social reproduction, and the urban. Yet what, conceptually, is at stake here? As for the urban, this is conventionally associated with a particular spatial-social form, 'the city', characterized by a range of phenomena, such as the development of increasingly dense and complex networks of physical infrastructures including roads, pipes, and wires.

Still in *The Urban Revolution* (2003 [1970]), Lefebvre famously argued that the traditional distinction between the city and the country no longer holds. Rather, the 'urban fabric' is rapidly extending across the globe, with rural areas transformed by the demands of urban centres. In recent years, a certain reading of this argument has inspired a number of provocative interventions within urban studies, according to which urbanization is now a 'planetary condition' (Brenner and Schmid 2012; Merrifield 2012; Brenner 2013, 2015). According to Brenner and Schmid, the key protagonists of the planetary urbanization thesis, no corner of the globe is left untouched by the process of capitalist urbanization. There is, it is suggested, no longer a rural or non-urban 'ontological Other'; what was previously conceived as such 'has been internalized into the very core of the urbanization process' (Brenner 2015, p. 174).

Yet planetary urbanization has been criticized as an overly totalizing account. For Ruddick et al. (2017), while it may be accurate to understand all aspects of the social totality as in some way connected to the urban process, myriad variegated experiences and struggles exist as not wholly subsumed by this. Indeed, Ruddick et al. (2017) argue that the planetary urbanization thesis obscures the role of the subjectivities and struggles of everyday life in shaping what they refer to as the 'social ontology' of the urban. Whereas planetary urbanization conjures up an encompassing capitalist urbanization process, within which counter-hegemonic praxis is reduced to empirical variation (Hart 2017), starting from the social ontology of the urban means understanding the urban process as in-part constituted through the contingent mediations of difference. Accordingly, the urban is rendered as 'undecidable' (Ruddick et al. 2017): as open to reconfiguration via the situated struggles of everyday life, including those struggles waged over social reproduction.

How, then, is social reproduction to be understood? Here, I follow the lead of the concrete struggles that inspire my argument. As I will show, APE activists see themselves as enacting a feminist praxis by virtue of their focus on struggles to 'sustain life' and, ultimately, their interest in crafting more collective and egalitarian ways of doing so. APE, then, are interested in what Mitchell, Marston, and Katz (2004) term 'life's work'. While, on this account, social reproduction has to do with 'the various ways in which life is made outside of work' (Mitchell, Marston, and Katz 2004, p. 23) – the spaces, practices, processes, and subjects implicated in the creation of value outside of waged labour – reproduction is never understood as sharply distinct from production but, instead, as dialectically related (see also Katz 2001b; Meehan and Strauss 2015). Thus, while APE's focus is on securing universal access to the urban infrastructures required for survival in the modern capitalist city – a struggle ostensibly enacted outside of the traditional 'workplace' – the concrete realities of those, like Esther, on the frontline of this battle, resist any kind of straightforward separation into binarized realms of work and non-work, productive and unproductive activity.

Sustaining life is, thus, an achievement built upon the reproduction 'both of the means of production and the labor power to make them work' (Katz 2001b, p. 711). While the home often takes centre stage in the analysis of this achievement, the materials and practices required for a household to reproduce itself are enmeshed within processes extending way beyond this. 'Life's work' is impossible without water and energy, for instance. In the modern capitalist city, these reproductive necessities are procured through connections to sprawling infrastructure networks such as water, gas, electricity, and transport, functioning within and between households, cities and, increasingly, national borders (Graham and Marvin 2001; Swyngedouw 2004; McFarlane and Rutherford 2008; Loftus 2012). While infrastructure is commonly understood as physical systems including the likes of cables, pipes, and roads, AbdouMaliq Simone (2004b) shows how, in Johannesburg, people themselves become infrastructure, with the reproduction of urban life in many parts of the world contingent upon informal social networks of improvized activity and collaboration. As already clarified, I understand infrastructure in this chapter as 'a platform providing for and reproducing life in the city' (Simone 2004b, p. 408) within which a variety of human and nonhuman practices and processes might be enrolled. In doing so, I recognize that the conceptual boundaries between social reproduction and infrastructure are rendered porous, with no clear line between practices and processes of reproductive labour and the infrastructures within which the former are enmeshed. What can be said is that the human infrastructures explored by Simone include the formal and informal social relationships, associations, and networks established in order to support social reproduction. APE itself, then, is a prime example of this kind of human infrastructure.

Returning to the relationship between social reproduction, the urban and infrastructure networks such as water and energy, the preliminary contention in this section has been that the urban is a process that is: i) both constitutive of and constituted by social reproduction; and that ii) renders social reproduction contingent upon an increasingly dense network of infrastructures – both physical and non-physical. In what follows, I ask what happens when, in a late capitalist city like Barcelona, a household lacks the funds required for inclusion within these infrastructure networks. How, in such instances, are human-infrastructure relations reconfigured, and with what implications for social reproduction and for the urban? And what, indeed, are the prospects for crafting less precarious, more socially just, forms of sociotechnical relations? It is to such questions that I will now turn, beginning with a contextualization of the Catalan study.

Contested Catalonia

The urban infrastructural struggles of APE emerge out of a specific historical geographical conjuncture. The financial crisis of 2008 had profound impacts in Spain, with subsequent rounds of government austerity measures resulting in widespread immizeration (García 2010; Charnock, Purcell, and Ribera-Fumaz 2014). Unemployment, for example, increased from 7.9% in the second quarter of 2007 to 25.8%

in the final quarter of 2012 (Instituto Nacional de Estadística 2015; Tirado Herrero and Jiménez Meneses 2016). As a result of stagnating incomes – alongside some of the sharpest electricity price hikes in Europe (Asociación de Ciencias Ambientales 2018) – people's ability to afford basic reproductive necessities such as energy and water has been curtailed: Spanish Household Budget Survey data indicates that in 2012, 17% of both Catalan and Spanish households spent over 10% of their income on domestic energy costs, almost triple the figure in 2007, at both the regional and national level (Tirado Herrero et al. 2014; Tirado Herrero and Jiménez Meneses 2016).

Crisis and austerity have been met with vociferous protest and resistance, largely concentrated in urban centres (García-Lamarca 2017). The Indignados uprising of 2011 was particularly influential, with squares in cities across the country occupied for months, demanding an end to austerity, calling out corruption amongst the Spanish political classes, and fleetingly enacting novel forms of participatory grassroots democracy (see Fominaya 2015). In the aftermath of their evictions, the radical energy of these urban occupations has filtered into an array of diverse struggles (Gonick 2016b). The anti-eviction movement la PAH, for example, formed in Barcelona in 2009, was bolstered by the events of 2011, going on to become one of the most celebrated urban social movements to have emerged in the wake of the 2008 crisis globally. Moreover, various so-called 'tides' in defence of public services have emerged, creating a sustained channel of grassroots resistance to government austerity. And, as will be discussed in more depth later, new progressive electoral projects have crystallized. This includes Podemos (We Can), which has become a major opposition party at the national level. And, in 2015, a series of 'municipalist' confluences that contested and won local elections across the country in 2015, with Barcelona En Comú's 'citizen's platform', led by former PAH spokeswoman Ada Colau (to become mayor of Barcelona), the most high profile of these victories (Eizaguirre, Pradel-Miquel, and García 2017).

This context of economic crisis and resistance has developed in tandem with an unfolding political crisis in Spain, which recently came to a head with the Catalan independence referendum of October 2017 – a referendum deemed unconstitutional by the Spanish state. Mass civil disobedience in defence of the right to Catalan self-determination was violently repressed by Madrid, culminating in the arrests of a number of pro-independence Catalan politicians (Jones 2017).

APE has no formal take on the question of independence, bringing together individuals with diverging perspectives on a question that, as I will soon suggest, is not seen as urgently relevant for many of those (though not all) within the network. Indeed, the broader context of crisis and resistance within Catalonia and its capital is highly relevant for under-standing the specific conditions under which a network like APE might emerge. In particular, the relationship to la PAH is crucial. La PAH's or-ganizing model, in which individual cases are fought through forms of collective solidarity-based support and direct action (see Gonick 2016a; García-Lamarca 2017), has formed the basis of APE's approach. Indeed, since its inception, APE has collaborated closely with la PAH, with both groups framing their struggles around the pursuit of the right to a digni-fied life, and many people participate fluidly across both networks – 'APE is PAH, PAH is APE', in the words of one activist I met.

#AguaParaEsther

As Sophie Gonick (2016a) has recently argued, la PAH are destabilizing the sanctity of property as a hegemonic institution of Spanish urban-ism through supporting households defaulting on mortgage repayments to remain in their homes (illegally). A lesser-known story is that with utilities firms reluctant to take on illegal occupiers as customers, the vast majority of these households are meeting their needs to heat, light, wash, drink, and cook in their homes through illegal reconnections to water, gas, and electricity networks.

This is currently the precarious lived reality of Esther, who was briefly introduced earlier in the chapter. Left unemployed in the aftermath of the financial crisis and without a means to pay her rent, Esther was forced to leave her home in April 2012, after which she has moved between street homelessness, short-term stays and, since 2013, illegally occupied housing. Upon moving into her current occupied dwelling in El Vendrell, Tarrago-na, with her teenage daughter, a complex struggle to access clean water to drink and wash with began. A nearby public water fountain provided a temporary solution, yet soon broke and, despite Esther's complaints, was left unfixed by the municipality. The nearest functioning public fountain that Esther could find was one situated 700 m away from her house down a steep hill. For months, Esther would make a daily trek up and down the hill to collect the household's water. Yet the heavy load soon took its toll, with the onset of severe back pain rendering this untenable. Another nearer fountain was then found, only to be cut-off by the municipality when they discovered Esther's reliance on it. Throughout this time, Es-ther had resolved not to resort to the option of what she termed 'stealing' water through illegally reconnecting the household to the city's formal water network. Yet her months of struggle left her with little other choice, and she therefore accepted the decision of a new housemate, who had the requisite technical knowledge, to proceed with establishing a tampered connection. This, though, was soon discovered by the household's water supplier, Aigües de Tomovi, who proceeded to cut off access.

It was at this point that Esther was introduced to APE, through a friend of a friend. Under the banner of #AguaParaEsther (water for Esther), the group resolved to pressure the municipality in El Vendrell – who had a majority 51% stake in Aigües de Tomovi – to use its position in the company to push for a regularized water connection for Esther's household. The campaign culminated in a three-month protest camp outside the municipality's offices between September and November 2015. This ultimately proved successful: the municipality granted APE's demand and the water supplier reconnected Esther to the water network, the first case of a formalized water connection for an illegally occupied dwelling in Spain, which has since set a precedent for many other cases to follow suit.

Esther, alongside other illegal occupiers, has had less success on the question of energy, with energy firms still refusing to take on illegal occupiers as customers. With no equivalent to a public fountain in electricity or gas, the only option here is illegal connections. Until mid-2016, Esther's tampered electricity connection had presented no problems. Yet in the summer of this year, a technician from the energy firm Endesa arrived to disconnect the household. In this first instance, Esther managed to persuade the technician to secretly reconnect her, once a photograph documenting the initial cut-off to Endesa management had been sent. Yet a series of similar visits have since occurred. And while, in some instances, other technicians have again empathized and reconnected Esther secretly, other times she has been less lucky and been left without electricity, leaving her with no option but to pay various informally operating electricians or else call upon the amateur know-how of friends to reconnect the house to the grid.

Esther's experience is typical of many of those I met within APE, whose membership largely comprises illegal occupiers in Barcelona's urban peripheries reliant upon irregular utilities connections.[3] My discussions with the protagonists of these everyday struggles to get by brought to mind the forms of creativity and improvization described by Simone (2004a, 2004b, 2009) with reference to the struggles of the urban poor across variegated cities within Africa and Southeast Asia. From Esther's backbreaking trek to the water fountain, to the operations of covert networks of informal electricians, and the forms of negotiation and co-operation between illegal occupiers and utility technicians, dominant dichotomies between the 'modern infrastructural ideal' of 'northern' cities (Graham and Marvin 2001) and the provisionality and irregularity of 'southern' cities are being exploded (Robinson 2006). Indeed, just as Baptista (2015, 2016), Silver (2016), and Pilo (2016, 2017) trace daily contestations around urban infrastructure metering, cut-offs, and tampered reconnections within Maputo, Accra, and Rio de Janeiro respectively, such 'incremental' struggles (Silver 2014) around water and energy access are also revealed as integral to the experience of the urban poor within the Catalan context.

In 'seeing from the south' (Roy 2009; Lawhon, Ernston, and Silver 2012; Parnell and Robinson 2012) and observing similarities between these northern and southern contexts, I do not intend to elide historical and geographical specificity, with the north/south binary a product of colonial capitalist histories (and presents) of violence and dispossession that must remain at the forefront of our theory-making. Nor, indeed, do I want to demarcate 'northern' and 'southern' cities as bounded units to be compared against each other. Rather, the form of relational comparison advocated by Hart (2017) and Loftus (2019) allows us to grasp how embodied practices, situated within distinct historical geographies, constitute and in turn are constituted by processes that – while articulating differently – connect urban life in Barcelona and elsewhere, destabilizing simplistic binaries between 'northern' and 'southern' cities, as well as dominant geographies of theory production.

The turn to the postcolonial context, in which the 'quiet encroachment' of subaltern groups (Bayat 2000) and hidden transcripts of everyday resistance (Scott 1990) are positioned at the heart of the political, helps in turn to expand thinking on the kinds of practices and spaces that constitute contested urban politics in a city like Barcelona. This, as alluded to already, is a city renowned for its spectacular outbursts of popular democracy, for example through collectivized workplace control during the Spanish Civil War; the Indignados uprising of 2011; and recent acts of mass civil disobedience around the question of Catalan independence. Yet the urban is not only contested in the workplace or the streets. Feminist urban geographers have long encouraged us to do away with divisions between public and private, or the politics of the street and the household, which still continue to plague urban studies (see, for instance, Hanson and Pratt 1988; McDowell 1999; Katz 2001b). This dichotomy, indeed, is revealed as spurious in the urban peripheries of Barcelona, home to a largely migrant working class whose everyday lives have become increasingly precarious (Butler 2004) and makeshift (Vasudevan 2015) in the face of financial crisis and austerity and who, I learnt during my time in Catalonia, have come to feel excluded from Catalan debates around sovereignty, independence, and nationhood. Here, urban politics begins with daily reproductive struggles to keep one's home safe, warm, and lit – struggles that spill over into the streets, disrupting public-private binaries. It is through the absence of regularized access to housing, water, electricity, and gas, that urban crisis is downloaded as a crisis of social reproduction; it is through engagements with irregular infrastructural connections, that new ways of navigating and producing the city (and urban subjectivities) are being performed.

I suggest that these shifting modalities of social reproduction are simultaneously fraught with insecurity and violence, and full of possibilities for more hopeful forms of urban life. To make this argument, next I reflect on the collective practices of APE in greater depth.

Feminist Praxis

APE, alongside la PAH, frame their struggles around the shared pursuit of the right to a dignified life. For APE, this begins with meeting basic reproductive needs. Their slogan, displayed on the group's signature red T-shirts alongside banners and signs, reads: 'Ni set, ni fred, ni foscor!' ('no thirst, no cold, no darkness!'). For APE activist Mònica, this approach is grounded in a feminist praxis:

> We didn't know that we would have so many women taking part no, so it's kind of a surprise, or it could seem random. But then you realise it's not, because the people who are sustaining these precarious lives are normally women inside the families ... Some feminist practices are present in APE, not only because we are many women there but because it's life-centred, like, sustaining lives is what matters. That is what guides our discourse, that is what guides our priorities maybe.

Here, the experiences of APE illustrate a core insight of social reproduction theory, namely, that a historically specific gendered division of labour under capitalism sees the artificial fetishization of production and reproduction, with the latter focusing on what Mònica terms 'sustaining precarious lives', coded as 'women's work' (Dalla Costa and James 1975; Mitchell, Marston, and Katz 2004; Meehan and Strauss 2015). A result is that the membership of APE is largely constituted by women. This, though, is clearly no guarantee of a feminist praxis. Rather, Mònica's point is that this feminist praxis emerges through the group's 'life-centred' perspective.

To be more specific, APE's praxis is feminist because it begins with the starting point of daily reproductive practices and, ultimately, seeks to transform these practices in liberatory ways. Clearly, a life-centred politics need not translate as a feminist politics. The daily reproductive struggles of particular social groups are often pitted against each other in the making of various forms of reactionary and exclusionary projects. The recent rise of the far right in Europe, for example, has in many contexts been premised upon endeavours to blame post-crisis immizeration on scapegoated racialized minorities. For APE, however, the focus is on a life-centred politics oriented around transcending the atomized and privatized relations of social reproduction in the neoliberal city to create more collective, just, and inclusive forms of urban life for all.

The departure point for this feminist praxis is the question of the reproductive relations internal to the community of people that constitute APE itself. For Federici (2012), struggles against capitalist enclosure and accumulation must begin by creating forms of mutually responsible community that allow for effective political organizing to be sustained over time. Indeed, feminist, queer, and anarchist perspectives have elevated 'prefigurative' politics to the centre of many contemporary struggles, encouraging attempts to craft new forms of socio-ecological relations in the here and now, rather than waiting until after some future revolutionary scenario (Holloway 2002; Maeckelbergh 2011; Ince 2012).

APE, then, seek to craft novel relations of collectivity and care within their own internal operations, seeking to socialize otherwise 'private' problems. Thus, individualized one-to-one models of traditional 'casework' are refused: when people come to the group seeking support, they are told that they must come to a fortnightly assembly, in which members of the network collectively offer advice and support with the goal of empowering people to find their own solutions. The opening statement read at the beginning of these assemblies powerfully presents the philosophy underpinning this approach:

> Welcome everyone, you are in the best place, you are no longer alone: this is a space of mutual support and trust. Be clear that this is not a crisis, it is a con. So firstly cover your basic needs (for instance food and clothing) and pay later. APE will never force you to do anything, all we will do is give you the tools to empower yourself. What do we want to achieve with these assemblies? A space of trust, collective support and mutual solidarity. Experience tells us that people will be concerned about other people's situations when they perceive that other people reciprocate concern for their own case. ... What we want is for people to know their rights and learn to defend themselves. We give you all of the tools but you are your own lawyer.

The intention, then, is to cultivate an ethos of solidarity rather than charity, with all encouraged to stay active in the group after their own cases have been addressed, in order to share the knowledge and hope they have gained with others in similar situations.

A frequently deployed tactic is 'accompaniment', in which an individual struggling with energy and water access will go to the office of the relevant utility firm alongside a group of APE activists, with the assembled mass negotiating on the case together and, often, refusing to leave until a satisfactory outcome is obtained. Alongside this, larger-scale protests and direct actions are organized – often occupations of utility firms' offices – which seek to win concessions on more generalized demands, the most important of which during my time in Catalonia, summer 2017, was the provision of regularized electricity connections for illegal occupiers.

Through these collective processes of assemblies, accompaniment and direct action, new reproductive relations are formed. The Alliance is referred to by its members as a family, in which efforts are made to ensure that all roles are respected and valued. 'The person who is in charge of bringing cookies, I think we owe this person everything,' Mònica told me. She continued: 'It makes the movement more inclusive and it … I think it also gives a lot of importance to all the work that is invisible, related to women.'

As with all prefigurative political projects, there is inevitably a degree of failure here, with dominant relations often re-inscribed (Pickerill and Chatterton 2006). First, as described above, APE operate within a gendered division of labour and, hence, the burden of struggle is disproportionately shouldered by women. That said, many men do participate in the group and, vitally, do so across a range of activities, rather than gravitating towards more typically 'macho' roles of strategic political visioning and militant street protest, as often occurs within the gendering of social movement work (Gonick 2016b). Class divisions within this new 'family' are however more pronounced, with a good deal of knowledge, expertise, and strategic power residing in the hands of a small group of university educated and experienced 'activists'. Still, there is a genuine commitment within the group to reflect upon and challenge the ways in which relations of domination are reproduced. For example, the culture of solidarity and mutual aid described above – a means of sharing leadership, expertise, and responsibility – begins to chip away at the aforementioned class divide. Indeed, the result is a group with far greater class and ethnic diversity than is, in my own experience, typical of many European urban activist networks.

Much of what is achieved through the formation of this new 'family' takes place at the affective level (Brown 2009; Brown and Pickerill 2009). In Esther's words: 'Talking about the Alliance makes me cry. For me it's a feeling, a really important thing.' 'What's special with the Alliance is the spirit, it comes from the soul. You really feel it,' for another APE activist named Pablo. 'You get involved, you care, you feel, you are allowed to cry,' as Mònica put it to me. Integral in this regard is the ethos of celebrating small victories. The group make a point of cherishing and publicizing every success, from a minor improvement on an individual's case through to a major concession from government or utilities firms, or simply just marking another year in the progression of the group. On this latter point, birthday parties are organized each year, on the date in which APE was first founded. Mònica commented:

> A birthday ... it's sacred, like, you need to celebrate that. And going there is as important as going to the mutual support meeting, no? And then you care about people, I mean, sharing food and singing, that's part of taking care, that's part of sustaining ourselves.

The point here is that victories, however minor they may seem, are an opportunity to feel powerful and hopeful and, in the process, to strengthen connections between activists and inspire optimism in further success. Indeed, at every demonstration of APE and la PAH, what has now become an internationally famous chant of 'Sí se puede!' ('Yes we can/We can win!') frequently erupts.

As part of this process of empowerment, APE conceives of their first task, upon the entrance of new members to the group, as combating the feelings of loneliness, fear, and shame associated with the absence of basics such as heat, light, and water (Sultana 2011). These feelings can be conceptualized as the affective dimension of the atomized and privatized regime of social reproduction that characterizes neoliberal urbanism. For Matt Huber (2013), this modality of social reproduction is conceptualized as 'hostile privatism', in which solidarity and collective responsibility give way to the 'entrepreneurial household', made to feel as though failing to secure the commodities necessary for reproducing itself is their own fault and responsibility. In the absence of support networks, this can leave people afraid about their futures.

The new 'family' fostered through the Alliance helps undermine hostile privatism and its accompanying affective register. People move from feeling afraid, alone, and ashamed to feeling supported, confident, and entitled to live with dignity. Lucia, a middle-aged Colombian migrant, told me:

> I've always been very rebellious, but I've never taken part in demonstrations, social activism, going to demonstrations, protesting against something, no ... Before I was afraid, since 2008 when the problem with the bank started and it was terrible ... In 2011 afraid didn't come close. It was frightening, I was terrified... But in 2013 I'm not afraid ... Because I'm able to fight with anybody. Banks, politicians, injustice. I'm not alone. There are many people with my same problem. You know when, for example, we say in the Alliance, when we do the welcome sessions, they say: 'you feel better when you see the other people have the same problems.' Now this is my strength.

The process of collective reproductive struggle, then, allows for possibilities open for the formation of new subjectivities. This is not a question of the emergence of a new essentialized or complete revolutionary consciousness, but rather the development of a dynamic and fraught process through which incomplete and fragmented subjects are perpetually made and remade (Haraway 1988; Loftus 2012).

Reproducing the Urban Otherwise

As APE prefigure more caring and collective reproductive relationships they, in turn, help foster new unruly subjectivities. Ruddick et al. (2017), in their endeavour to articulate a social ontology of the urban, argue that it is in the messy and fraught contestations of everyday life that subjects, and hence the openings and closures for radical politics, are made and remade. Thinking in this vein, I suggest that the processes of subjectivation underway within APE open up possibilities for more emancipatory forms of urban life.

First, APE's feminist praxis – premised upon the creation of caring and collectivized human connections otherwise made irregular by hostile privatism challenges the commodification of social reproduction. As a step towards this, APE's campaigning forced a new Catalan law to be introduced in 2015, illegalizing water, electricity, and gas disconnections for people designated as 'vulnerable'. While this has more or less eradicated cut-offs for those who own or rent their homes, utilities firms have found a legislative loophole to allow disconnections for those living in illegally occupied homes, thus causing the situation of Esther and others described previously. Throughout my research with APE, their focus was on winning the right to regularized connections for illegal occupiers. This fight is tied to the demand for debt to be written off, with all costs covered by utilities firms. This portends a situation in which illegal occupiers could gain safe regularized connections, without a monetary cost. The result is that the commoditization of the necessities of social reproduction is challenged, with the political imagination expanded such that a world beyond the commodity form becomes easier to imagine.

This feminist praxis, moreover, has constituted the foundations of one of the most ambitious attempts to remake and reclaim the city that has been seen in Europe in recent years. This is the 'municipalist' project of Barcelona En Comú, a so-called citizen's platform bringing together diverse grassroots initiatives and formally fringe progressive parties, which in May 2015 won minority control of the city council in the Catalan capital, less than one year after its formation. Barcelona en Comú was founded by activists involved in the city's social movements, including

APE, without any previous political party experience. Their agenda (see Barcelona En Comú 2016; Eizaguirre, Pradel-Miquel, and García 2017) is premised upon, first, the defence of 'social rights', including the right to safe regularized water and electricity connections, and other necessities of social reproduction. Second, Barcelona En Comú seeks to enact a 'different way of doing politics', in which the institutions of the local state become a vehicle to facilitate direct democracy and self-governance. Neighbourhood assemblies and online democratic tools are used to shape policy, and the council are attempting to promote increased popular participation in the economy through the re-municipalization of services and utilities and the expansion of co-operatives and mutuals.[4]

The feminist praxis of APE has been important in the formation of Barcelona En Comú in two connected ways. First, the dedicated grassroots organizing of APE and other urban movements in the city provided a pre-existing infrastructure through which a municipalist electoral project could quickly be built and gain momentum. In the words of Kate Shea Baird, a member of Barcelona En Comú's International Working Group and Coordinating Committee: 'most of the important work was already done' prior to the decision to contest the elections. From networks of activists ready to campaign, strong channels of communication and relationships between these networks and the city's neighbourhoods, and the crystallization of a common sense within the city around the key tenets of Barcelona En Comú's 'crowdsourced' manifesto, the conditions of possibility for a successful progressive electoral venture had been established. Without the previous endeavours of groups such as APE and la PAH – endeavours embedded within daily reproductive practices – the municipalist project could not have got off the ground.

Second, the feminist approaches of APE and la PAH have informed Barcelona En Comu's commitment to what they term 'the feminization of politics', which pursues a mode of politics grounded in relationships and everyday life. As Kate Shea Baird and fellow Barcelona En Comú activist Laura Roth (2017, n.p.) put it in a recent blog post:

> ... a feminized politics seeks to emphasize the importance of the small, the relational, the everyday, challenging the artificial division between the personal and the political. This is how we can change the underlying dynamics of the system and construct emancipatory alternatives.

The idea is to reject binaries between political ends and means, refusing the panicked urgency of attempts to craft revolution 'by any means necessary' in favour of a more patient emphasis on the importance of caring and egalitarian relationships in the immediate conjuncture – this emphasis emerging not out of an essentialized conception of the 'caring woman' but, rather, out of an understanding of the value of care grasped through the process of feminist struggle. Here, we see the prefigurative politics enacted by the likes of APE deeply influencing the ways in which Barcelona En Comú are endeavouring to create urban commons.

The extent to which Barcelona En Comú (still in government at the time of writing, albeit with one fewer councilors in power after the 2019 municipal elections) are succeeding in their attempts to feminize politics, defend social rights, and democratize the city is debatable. Successes claimed by the initiative include significant reductions in evictions and utilities cut-offs, alongside the formation of municipal funeral and energy companies. Yet the initiative has, thus far, been unable to proceed with a number of flagship manifesto pledges including re-municipalizing water and closing down a nearby immigration detention centre. An in-depth assessment of the project is beyond the scope of this chapter. My contention here is simply that to the extent that Barcelona En Comú are succeeding in remaking the city, the role of groups like APE, whose praxis begins at the point of daily reproductive practices, should be recognized as vital.

Conclusion

Austerity urbanism in Catalonia, and subsequent modifications to socio-technical relationships between people and infrastructure, have heightened the fragility and violence that characterizes social reproduction under capitalism. Yet the conclusion I want to draw from the discussion of this chapter is that urban processes do not simply flow downwards to shape reproductive practices. Rather, the materialities, affects, and subjectivities of social reproduction are themselves constitutive of the urban process itself. Thus, while urban restructuring will always reconfigure reproductive relations, we must simultaneously recognize that by reproducing ourselves differently, we might open possibilities for remaking the city. It is in this sense that a co-evolving dialectical relationship between social reproduction and the urban can be claimed. Infrastructure networks provide one materiality through which this relationship is constituted. When urban transformation reconfigures human-infrastructure relations, reproductive practices shift. When new forms of human-infrastructure relations are built in the reconfiguration of reproductive practices, the urban process is remade.

In making this argument, I hope to have made clear the political stakes of a feminist outlook that centres social reproduction. The feminist praxis of APE, I have argued, is grounded in the reproductive labour necessary to sustain life and, more specifically, in the rendering of this labour as a more collective, caring, and liberatory endeavour. My suggestion has been that from the foundations of this prefigurative politics, emancipatory transformations of the urban process may become possible.

Addressing the closing plenary of 'Fearless Cities', an international gathering of municipalist projects held in June 2017, Barcelona En Comú leader and Mayor of Barcelona Ada Colau concluded: 'From our fragility, we affirm that we are not afraid because we are together.' Fearless cities remain fragile cities. The point is not to deny our precarity but, rather, to recognize this as a shared ontological condition out of which a powerful collective politics could be forged (Butler 2004). One way in which urban life in the modern capitalist city is produced as precarious is the rendering of our social reproduction as contingent upon commodified access to the human and non-human infrastructures necessary to sustain life's work. APE enacts a collective infrastructural politics that recognizes this fragility and from this, builds new forms of togetherness. From this fearless reproductive politics, fearless cities might just be won.

Acknowledgements

My utmost gratitude, first, to all of the new friends and comrades in APE who generously welcomed me to their group and inspired the chapter. Particular thanks in this regard are due to Mònica Guiteras for her invaluable assistance throughout my time in Catalonia. Many thanks also to Alex Loftus, Luis Andueza, Archie Davies, Pratik Mishra, Hannah Schling, Sergio Tirado Herrero, Hyerim Yoon, and three editors, Linda Peake, Rajyashree Reddy, and Gökbörü Sarp Tanyildiz for their constructive criticisms and feedback on the chapter and its previous iterations.

Notes

1 The names of some research participants have been changed in the instances where anonymity was chosen.

2 I oscillate between the Castillian and Catalan languages throughout the chapter, attempting to retain fidelity to the terminology I encountered in Catalonia.

3 While a recent report commissioned for Barcelona city council suggests that pinchazo connections are a relatively fringe practice (Tirado Herrero 2018), Esther's situation is similar to many of the afectadas I met within APE. Those who seek support through APE tend to be enduring the most severe forms of energy poverty and, thus, pinchazo connections become an important last resort.

4 'Re-municipalisation' refers to the process of privatization being reversed through the transferral of ownership and management of services and utilities to municipal institutions (see Angel 2017).

References

Angel, J. (2017). Towards an energy politics in-against-and-beyond the state: Berlin's struggle for energy democracy. *Antipode* 49 (3): 557–576.

Asociación de Ciencias Ambientales (2018). *Pobreza Energética En España: Hacia Un Sistema de Indicadores Y Una Estrategia de Actuación Estatal.* Madrid: Asociación de Ciencias Ambientales.

Autonomous Geographies Collective (2010). Beyond scholar activism: Making strategic interventions inside and outside the neoliberal university. *ACME* 9 (2): 245–275.

Baptista, I. (2015). 'We live on estimates': Everyday practices of prepaid electricity and the urban condition in Maputo, Mozambique. *International Journal of Urban and Regional Research* 39 (5): 1004–1019.

Baptista, I. (2016). Maputo: Fluid flows of power and electricity – Prepayment as mediator of state-society relationships. In: *Energy Power and Protest on the Urban Grid: Geographies of the Electric City* (eds. A. Luque-Ayala and J. Silver), 112–131. London: Routledge.

Barcelona En Comú. (2016). *How to win back the city En Comú: Guide to building a citizen municipal platform.* https://barcelonaencomu.cat/sites/default/files/win-the-city-guide.pdf (accessed 9 October 2020).

Bayat, A. (2000). From 'dangerous classes' to 'quiet rebels'. *International Sociology* 15 (3): 533–557.

Brenner, N. (2013). Theses on urbanization. *Public Culture* 25: 85–114.

Brenner, N. (2015). Towards a new epistemology of the urban? *City* 19 (2–3): 151–182.

Brenner, N. and Schmid, C. (2012). Planetary urbanization. In: *Urban Constellations* (ed. M. Gandy), 10–13. Berlin: Jovis.

Bridge, G., Bouzarovski, S., Bradshaw, M. et al. (2013). Geographies of energy transition: Space, place and the low-carbon economy. *Energy Policy* 53: 331–340.

Brown, G. (2009). Space for emotion in the spaces of activism. *Emotion, Space and Society* 2 (1): 24–35.

Brown, G. and Pickerill, J. (2009). Editorial: Activism and emotional sustainability. *Emotion, Space and Society* 2 (1): 1–3.

Butler, J. (2004). *Precarious Life: The Powers of Mourning and Violence.* London: Verso.

Charnock, G., Purcell, T., and Ribera-Fumaz, R. (2014). *The Limits to Capital in Spain: Crisis and Revolt in the European South.* New York: Palgrave Macmillan.

Chatterton, P., Featherstone, D., and Routledge, P. (2013). Articulating climate justice in Copenhagen: Antagonism, the commons, and solidarity. *Antipode* 45 (3): 602–620.

Dalla Costa, M. and James, S. (1975). *Power of Women and the Subversion of the Community*. Bristol: Falling Wall Press.

Derickson, K.D. and Routledge, P. (2015). Resourcing scholar-activism: Collaboration, transformation, and the production of knowledge. *The Professional Geographer* 67 (1): 1–7.

Eizaguirre, S., Pradel-Miquel, M., and García, M. (2017). Citizenship practices and democratic governance: 'Barcelona En Comú' as an urban citizenship confluence promoting a new policy agenda. *Citizenship Studies* 21 (4): 425–439.

Federici, S. (2012). *Revolution at Point Zero*. Oakland, CA: PM Press.

Fominaya, C.F. (2015). Debunking spontaneity: Spain's 15-M/Indignados as autonomous movement. *Social Movement Studies* 14 (2): 142–163.

García, M. (2010). The breakdown of the Spanish urban growth model: Social and territorial effects of the global crisis. *International Journal of Urban and Regional Research* 34 (4): 967–980.

García-Lamarca, M. (2017). From occupying plazas to recuperating housing: Insurgent practices in Spain. *International Journal of Urban and Regional Research* 41 (1): 37–53.

Gonick, S. (2016a). From occupation to recuperation: Property, politics and provincialization in contemporary Madrid. *International Journal of Urban and Regional Research* 40 (4): 833–848.

Gonick, S. (2016b). Indignation and inclusion: Activism, difference, and emergent urban politics in postcrash Madrid. *Environment and Planning D: Society and Space* 34 (2): 209–226.

Graham, S. and Marvin, S. (2001). *Splintering Urbanism: Networked Infrastructures, Technological Mobilities and the Urban Condition*. London: Routledge.

Halvorsen, S. (2015). Militant research against-and-beyond itself: Critical perspectives from the university and Occupy London. *Are* 47 (4): 466–472.

Hanson, S. and Pratt, G. (1988). Reconceptualizing the links between home and work in urban geography. *Economic Geography* 64 (4): 299–321.

Haraway, D. (1988). Situated knowleges: The science question in feminism and the privilege of partial perspectives. *Feminist Studies* 14 (3): 575–599.

Hart, G. (2017). Relational comparison revisited: Marxist postcolonial geographies in practice. *Progress in Human Geography* Online. doi:10.1177/0309132516681388

Holloway, J. (2002). *Change the World without Taking Power*. London: Pluto.

Huber, M. (2013). *Lifeblood: Oil, Freedom, and the Forces of Capital*. Minneapolis, MN: University of Minnesota Press.

Ince, A. (2012). In the shell of the old: Anarchist geographies of territorialisation. *Antipode* 44 (5): 1645–1666.

Instituto Nacional de Estadística. (2015). Tasas de paro por sexo y grupo de edad. *Instituto Nacional de Estadística.* www.ine.es/jaxiT3/Tabla. htm?t=4086 (accessed 9 October 2020).

Jones, S. (2017). Spanish judge jails eight members of deposed Catalan government. *The Guardian* (2 November). www.theguardian.com/world/2017/nov/02/spanish-court-question-catalonia-separatists-except-puigdemont (accessed 9 October 2020).

Katz, C. (2001a). On the grounds of globalization: A topography for feminist political engagement. *Signs* 26 (4): 1213–1234.

Katz, C. (2001b). Vagabond capitalism and the necessity of social reproduction. *Antipode* 33 (4): 709–728.

Lawhon, M., Ernston, H., and Silver, J. (2012). Provincializing urban political ecology: Towards a situated UPE through African urbanism. *Antipode* 46 (2): 497–516.

Lefebvre, H. (2003 [1970]). *The Urban Revolution* (trans. R. Bononno). Minneapolis, MN: University of Minnesota Press.

Loftus, A. (2012). *Everyday Environmentalism: Creating an Urban Political Ecology.* Minneapolis, MN: University of Minnesota Press.

Loftus, A. (2019). Political ecology I: Where is political ecology? *Progress in Human Geography* 43 (1): 172–182. doi:10.1177/0309132517734338

Maeckelbergh, M. (2011). Doing is believing: Prefiguration as strategic practice in the alterglobalization movement. *Social Movement Studies* 10 (1): 1–20.

McDowell, L. (1999). *Gender, Identity and Place: Understanding Feminist Geographies.* Minneapolis, MN: University of Minnesota Press.

McFarlane, C. and Rutherford, J. (2008). Politicised infrastructures: Governing and experiencing the fabric of the city. *International Journal of Urban and Regional Research* 32 (2): 363–374.

Meehan, K. and Strauss, K. eds. (2015). *Precarious Worlds: Contested Geographies of Social Reproduction.* Athens, GA: The University of Georgia Press.

Merrifield, A. (2012). The urban question under planetary urbanization. *International Journal of Urban and Regional Research* 37 (3): 909–922.

Mitchell, K., Marston, K., and Katz, C. eds. (2004). *Life's Work: Geographies of Social Reproduction.* Oxford, CA: Blackwell.

Nagar, R. (2002). Footloose researchers, 'traveling' theories, and the politics of transnational feminist praxis. *Gender, Place & Culture* 9 (2): 179–186.

Parnell, S. and Robinson, J. (2012). (Re)theorizing cities from the Global South: Looking beyond neoliberalism. *Urban Geography* 33 (4): 593–617.

Pickerill, J. and Chatterton, P. (2006). Notes towards autonomous geographies: Creation, resistance, and self-management as survival tactics. *Progress in Human Geography* 30 (6): 730–746.

Pilo, F. (2016). Rio de Janeiro: Regularising favelas, energy consumption and the making of consumers into customers. In: *Energy Power and Protest on the Urban Grid: Geographies of the Electric City* (eds. A. Luque-Ayala and J. Silver), 67–85. London: Routledge.

Pilo, F. (2017). A socio-technical perspective to the right to the city: Regularizing electricity access in Rio de Janeiro's favelas. *International Journal of Urban and Regional Research* 41 (3): 396–413.

Robinson, J. (2006). *Ordinary Cities: Between Modernity and Development.* London: Routledge.

Roth, L. and Shea Baird, K. (2017). Municipalism and the feminization of politics. *Roar.* https://roarmag.org/magazine/municipalism-feminization-urban-politics (accessed 9 October 2020).

Roy, A. (2009). The 21st-century metropolis: New geographies of theory. *Regional Studies* 43 (6): 819–830.

Ruddick, S., Peake, K., Tanyildiz, G.S. et al. (2017). Planetary urbanization: An urban theory for our time? *Environment and Planning D: Society and Space* Online First: 1–18.

Scott, J.C. (1990). *Domination and the Arts of Resistance: Hidden Transcripts.* New Haven, CT and London: Yale University Press.

Silver, J. (2014). Incremental infrastructures: Material improvisation and social collaboration across post-colonial Accra. *Urban Geography* 35 (6): 788–604.

Silver, J. (2016). Disrupted infrastructures: An urban political ecology of interrupted electricity in Accra. *International Journal of Urban and Regional Research* 39 (5): 984–1003.

Simone, A. (2004a). *For the City yet to Come: Urban Life in Four African Cities.* London: Duke University Press.

Simone, A. (2004b). People as infrastructure: Intersecting fragments in Johannesburg. *Public Culture* 16: 407–429.

Simone, A. (2009). *City Life from Jakarta to Dakar: Movements at the Crossroads.* New York: Routledge.

Sultana, F. (2011). Suffering for water, suffering from water: Emotional geographies of resource access, control and conflict. *Geoforum* 42 (2): 163–172.

Swyngedouw, E. (2004). *Social Power and the Urbanization of Water: Flows of Power.* Oxford, CA: Oxford University Press.

Tirado Herrero, S. (2018). *Indicadores municipales de pobreza energética en la Ciudad de Barcelona.* Barcelona: RMIT Europe. https://habitatge.barcelona/sites/default/files/documents/indicadors-municipals-de-pobresa-energetica-a-la-ciutat-de-barcelona.pdf (accessed 9 October 2020).

Tirado Herrero, S. and Jiménez Meneses, L. (2016). Energy poverty, crisis and austerity in Spain. *People, Place and Policy* 10 (1): 42–56.

Tirado Herrero, S., Jiménez Meneses, L., López Fernández, J.L. et al. (2014). *Pobreza energética en España. Análisis de tendencias.* Madrid: Asociación de Ciencias Ambientales.

Vasudevan, A. (2015). The makeshift city: Towards a global geography of squatting. *Progress in Human Geography* 39 (3): 338–359.

Weeks, K. (2011). *The Problem with Work: Feminism, Marxism, Antiwork Politics, and Postwork Imaginaries.* Durham, NC: Duke University Press.

6

Global Restructuring of Social Reproduction and Its Invisible Work in Urban Revitalization

Faranak Miraftab (University of Illinois, Urbana-Champaign)

Introduction

Social reproduction is key to processes of production and accumulation, and it is central to making and unmaking urban spaces. The ability of patriarchal and (neo)colonial relations to render this central condition invisible is fundamental to the system of capitalism. Under capitalism, racialized and gendered systems of sociocultural hierarchies and norms make social reproduction work invisible (Mies 2007; Federici 2014). Feminist urban scholars have pointed out how this invisibilization is achieved through reorganizing spatial relations – a process that allows capitalism to temporarily resolve its crises of accumulation. The particularities of how patriarchal gendered normalization facilitates the spatial fix, and the spatial and temporal conditions under which social reproduction work is gendered, racialized, and made invisible, play out differently in different times and places. Thus, the agenda of an anticolonial, antiracist feminist urban scholarship must be to make visible the variations and particularities in such processes and relationships.

This chapter concerns the intertwining processes of transnational social reproduction and urban revitalization of the Rustbelt – the term euphemistically used in reference to the industrial regions of the United States

A Feminist Urban Theory for our Time: Rethinking Social Reproduction and the Urban, First Edition. Edited by Linda Peake, Elsa Koleth, Gökbörü Sarp Tanyildiz, Rajyashree N. Reddy & darren patrick/dp.
© 2021 John Wiley & Sons Ltd. Published 2021 by John Wiley & Sons Ltd.

after manufacturing capital started leaving in the 1960s for locations with cheaper labour and higher profit margins.[1] Abandoned by manufacturing firms, these places set in course a process of depopulation and real estate downturn; some previously vigorous urban centres like Detroit, with loss of jobs and population, became a shadow of their previous selves and others became boarded-up ghost towns. Over the last three decades, however, the de-populated small towns of the Rustbelt have started to see a wave of new life in the arrival of globally displaced and marginalized populations. The displaced people who inhabit these left-behind spaces make them their new grounds of life making and revitalize them – not only socioeconomically but also physically, as in place-making or in-placement. The remaking of abandoned spaces into places composed of homes, schools, thriving markets, and factories requires a tremendous amount of socially reproductive labour, and that labour is distributed across multiple sites, not only locally but also transnationally. I focus on the ongoing transnational social reproduction processes that are constitutive of this process, yet are rendered invisible and thus unaccounted for. Metro-centric urban scholarship does not pay attention to the socio-spatial dynamics of globalization and change in these small towns and politicians write them off their books as political backwaters. Ignored by academic research and misunderstood by politicians, these left-behind spaces have become the Trumplands of the US political landscape and breeding grounds for the nativist discourse amongst locals that feel abandoned, not only by employers but also by politicians and public media. This chapter, by focusing on processes of urban revitalization due to new flows of globally displaced labour, contributes on two fronts: on the one hand, to critical transnational urban scholarship showing how even the small towns labelled as backwaters of globalization are intensely nestled in global processes of labour displacement; and on the other, to debunking the deafening nativist discourse of White nationalism which claims new migrant workers siphon off their local resources through remittances sent to their communities of origin. The contribution made by transnational families, that bring new inhabitants to small but intensely industrial and globally connected spaces of the Rustbelt, contributes not only to global urban scholarship by making the dynamics of small towns visible, but also to the political agenda of anticolonial and antiracist struggle.

I start the chapter with a discussion of the landscape of new inequalities in the Rustbelt, then briefly review the insights we have gained through what is now classic feminist urban scholarship on how patriarchal gendered normalization facilitates the spatial fix that took place through postwar suburbanization in America and through global austerity policies with devastating effects for subordinate communities – what I refer to as global South communities. Building on these insights, I move on to a more detailed discussion of how social reproduction comes to serve crises of capitalism in this latest era of global capitalism. Focusing on social and spatial redevelopment of the Rustbelt, in particular its much ignored small towns, I discuss what I theorize as a global restructuring of social reproduction, and I show how social reproduction work performed by local and transnational families is made invisible, not only through its gendered normalization but also through its spatial fragmentation both across the globe and within existing postcolonial racialized urban hierarchies.

A clarifying note here, with respect to the Rustbelt and its varied spaces, metropolitan and beyond. I refer to small towns in the Rustbelt as non-metropolitan areas, which is not to be interpreted in some traditional sense as if these are 'non-urban' areas.[2] Small towns in the US have always been places of flows, of in- and out-migration, of displacement (of Native Americans, for example), and of in-placement. Emergent urban scholarship, however, with its metrocentric theorization, silences a range of places and place-based politics, such as those I discuss in relation to the Rustbelt's small towns.

A Landscape of New Inequalities in the Rustbelt and Its Social and Spatial Transformation

Wilson and Miraftab (2015) wrote about the ways in which the landscape of the Rustbelt, with its abandoned spaces 'left behind' by the restructuring of the global capitalist economy and its attending different regimes of accumulation, now offers a new round of opportunities for accumulation through two types of economies: the place-based and the parasitic. Place-based economies, which cannot function outside their current locations and must bring in cheaper labour from outside to reduce costs, include service jobs or jobs in food and agribusiness, such as animal slaughtering and meatpacking. Once the main industries of a region have left in pursuit of cheap labour forces, the industries that are place-bound and not able to move have stayed put but enjoyed cheap labour that has crossed borders to reach them. These low-waged new arrivals in the Rustbelt are criminalized and marginalized, as 'illegal aliens' who 'cannot be trusted' and 'are up to no good', a nativist narrative that forces them to accept wages lower than what their US-born counterparts would have accepted. The economic vulnerabilities of the new workers in these areas offer another opportunity to global capital through its parasitic economies such as those of check cashers,[3] payday lenders,[4] pawnshops, temporary labour agencies,[5] and day labour sites, which move their facilities anywhere that low-waged and vulnerable workers can be found. These institutions of the parasitic economy should not be regarded as 'mom-and-pop' twilight economies, as they often were for earlier generations of immigrants. Instead, these stores are often outlets for national or multinational chains such as Western Union, Segue, and Money Express. Western Union, for example, operates across America, Europe, and the global South and is enriched by capital from the Rockefeller Trust and Union National Bank (Karger 2005).

Like vultures, various forms of capital return to feast in the left-behind spaces in another round of exploitation and devastation. While footloose manufacturing firms left, place-based industries, such as agro-industries, moved their plants closer to their raw material and parasitic economies connected with financial capital, such as check cashers and payday lenders, found lucrative opportunities in these left behind spaces to feed off those who could not leave and those who arrived predominantly as displaced populations (a more accurate reference than immigrant), and who are marginalized, racialized, and criminalized as workers. Despite this grim picture, many of the new arrivals, who have been displaced from their homes in their communities of origin, have made new homes in the small towns of the Rustbelt.

This chapter situates social reproduction within the processes of revitalization in the small towns of the Rustbelt at this moment of late capitalism. I ask: What are the processes and practices of social reproduction underlying the small-town urban revitalization in the Rustbelt and how are they made invisible? To answer this question, I draw on a body of the multi-sited ethnographic work I conducted between 2006 and 2012, which reveals the invisible work of transnational social reproduction involved in sustaining low-waged labourers in the Rustbelt and for the Rustbelt's revitalization. I use the example of Beardstown, Illinois, a meatpacking town, and its revitalization following the recruitment of Mexican and West African workers to the local meatpacking plant. After the flight of most industry from this town during the 1970s and 1980s, just one industry, which slaughters, processes, and packs pork, remains in business. Following the industry's restructuring in the mid-1980s, the Cargill meat-production facility cut its workers' wages by 2.5 dollars per hour, prompting the departure of a large number of its workers from the plant and the town. In light of the high labour turnover rate, the company turned to recruitment of immigrant workers – first, in the early 1990s, amongst Spanish-speaking Central Americans and Mexicans and then, in the early 2000s, amongst French-speaking West Africans (at the time predominantly Togolese). By 2010, this process had halted the depopulation, keeping the number of inhabitants stable at around 6000; it had also shifted the demographics of the town from 99% White in 1990 to 61% White in 2010. In a matter of ten years – between 1990 and 2000 – the number of people of colour living in Beardstown grew by an astonishing factor of 32. By 2010, when US Census staff were in Beardstown, only three out of every five residents identified themselves as being non-Hispanic White, and one out of every five was born outside the United States (Miraftab 2016, p. 4). Consequently, the town was given a substantially new life as newcomers fixed up abandoned, boarded-up houses, began buying at local grocery stores and gas stations, and started signing up their children at the public schools. In short, they revitalized the Rustbelt (see Grey and Woodrick 2005; Griffith 2008; Miraftab 2012, 2016; Sandoval and Maldonado 2012).

The case of Beardstown is not an exception. It is part of a larger trend documented in the United States by a large body of research. The depopulated communities of the Rustbelt, where industries have been leaving since the 1960s in search of destinations with lower wages, have now become the home of new populations. These are people[6] who have been dispossessed of their livelihood and displaced from their home communities to wherever they can find a livelihood for themselves and their families, a process Sassen (2014) calls expulsion. Families of these displaced workers might have accompanied them in this process or been left behind in places of origin. This movement into the Rustbelt is widely documented by demographers, who have shown that, since the 1990s, the percentage of foreign-born populations has been decreasing in the traditional gateway metropolitan areas of Los Angeles, Miami, Chicago, New York, and Houston, while simultaneously increasing in non-gateway areas (Gozdziak and Martin 2005; Frey 2006; Portes and Rumbaut 2006). Data shows that the percentage of the total foreign-born population in the United States living in the traditional gateway metropolitan areas shrank from 43% in 1990 to 33% in 2010, while the percentage of foreign-born residents living in non-gateway destinations grew from 56 to 66% (calculated from IPUMS data; see Ruggles et al. 2010). In the last two decades, non-metropolitan counties of the Rustbelt, which previously had little, if any, foreign-born population, have seen the end or even the reversal of their population shrinkage thanks to the arrival of a new and growing foreign-born labour force (Durand, Massey, and Capoferro 2005). Apart from the lower cost of living, displaced peoples arrive in these areas in search of employment, which is often offered to them by manufacturing sectors that need to stay close to their raw materials – namely agriculture and animals. These industries, however, in order to maintain or increase profitability in the face of global competition, offer depressed wages that are unattractive to the local labour force (Stull, Broadway, and Griffith 1995).

In *Global Heartland* (2016), I have written extensively about the complex local transnational processes and practices that account for this social and spatial transformation of the Rustbelt and particularly the revitalization of the meatpacking town described above. Here, I share one of the arguments made in the book to unfold how the social reproduction work performed locally and trans-locally within and across national borders is made invisible in the processes of Rustbelt revitalization. Specifically, I articulate the strategies that I observed amongst displaced workers and their transnational families, which I call the *global restructuring of social reproduction*, and explain how this process – invisible to the host communities – contributes to urban revitalization and development.

Social Reproduction and Its Global Restructuring

In its most prominent and common definition, social reproduction has come to indicate biological reproduction, reproduction of the labour force, and the individual and institutional provision of care (Hopkins 2015). Of course, this biological reproduction of the next generation of workers requires not only food, clothing, and shelter, but also collective services such as sanitation, water, sewage, schools, and transport – what Castells (1983) called collective consumption. I use the term *social reproduction* in a broad sense, to include the reproduction of place, cultural identities, traditions, and sense of pride. In this broad interpretation, social reproduction is not limited to biophysical reproduction and is not driven merely by an economic logic but also accounts for the emotional and cultural logics that are important in workers' emergent social reproduction strategies and practices. It includes the individual, family, collective, and institutional care that is needed to bring workers to the workplace and reproduce the next generation of labourers, who in turn create the wealth of society – namely, a comprehensive understanding of social reproduction that Arruzza, Bhattacharya, and Fraser (2019) call 'life making' and Katz (2001) refer to the 'messy and fleshy' stuff of everyday life that allows people and their labour power to be reproduced under capitalism (also see Bhattacharya 2017). In the discussions of social reproduction amongst Marxists and Marxist-feminists, and for most of the 20th century, production and social reproduction were conceptually and analytically treated separately, and spatially were assumed to take place in proximity to one another (for a critique of this position and the blurriness of the production and reproduction in life's work, see Mitchell, Marston, and Katz 2004). Today, however, no such assumption can be made where production happens at home (Miraftab 1996; Prugl 1999) and workers in one location have their families cared for elsewhere (Hontagneu-Sotelo and Avila 1997; Hochschild 2000; Parreñas 2001; Pratt 2005; Herrera 2008). Moreover, workers in one location may be making places and developing communities across the world and in multiple locations – for example, for immigrant and displaced workers social reproduction and place-making may take place both in communities of origin and at their destination (see Sandoval and Maldonado 2012; Sarmiento and Beard 2013; Irazábal 2014).

Within such an analytic framework of social reproduction, I pay attention to trans-local and transnational processes and practices of social reproduction, not only at household but also at the collective level of local communities and urban areas, in response not only to biophysical but also emotional and cultural needs that may or may not be driven through an economic logic. In particular, I focus on processes of social reproduction when production and social reproduction do not take place in proximity, but around the world and across international borders. I theorize such processes as the global restructuring of social reproduction.

In the current era of global capitalism, the global restructuring of social reproduction refers to processes that socially, temporally, and spatially reorganize people's biophysical, social, and cultural reproduction responsibilities within and across national borders. I discuss this in terms of outsourcing the care work for workers' immediate family members – for example, the raising of younger children or caring for elders in communities of origin. Such outsourcing may also result in fragmentation of the life cycle, whereby the beginning and the end of life (usually the downtimes of one's life based on labour market participation) takes place in communities of origin supported by families and institutions there, while the most economically productive segment takes place, under exploitative conditions, in the US labour market.

The restructuring of social reproduction has profound implications for place-making. Feminist urban scholarship has revealed how social reproduction and normalization of gender roles and expectations have come to the rescue of capital in its previous rounds of crises and the needed spatial fixes, be it restructuring of urban space (suburbanization) or restructuring of basic service provision in urban space (re-privatization). Here I focus on the invisible transnational social reproduction work that develops Rustbelt communities of the US heartland. I highlight how displaced workers' transnational social reproduction involves place-making in their places of origin and destinations. Before further discussion, allow me a few words on the methodology and approach.

Relational Framing and Radical Feminist Urban Scholarship

How we tell a story matters. Our framing of the story determines what we reveal and what we obscure; what we place at the centre and what remains at the margin; and what defines the structure or becomes a marginal element in the construction of the story. The story of a revitalized packing town in the Rustbelt framed merely through the processes of immigration to the heartland reveals some aspects of the global processes and obscures others. On the one hand, it reveals how international migration of the labour force transforms local communities, which, despite their geographical isolation, are intimately involved in the production of global capital and its processes of accumulation. On the other hand, it renders invisible the stories of dispossession and displacement that produce a migrant labour force in the first place. By and large, most work on meatpacking towns and heartland revitalization takes a local perspective (see, for example, Stull, Broadway, and Griffith 1995; Fennelly and Leitner 2002; Grey and Woodrick 2005). Similarly, it might reveal the restructuring of production processes at the global scale and the restructuring of the meatpacking industry, while invisibilizing the processes of social reproduction performed from afar in labourers' communities of origin. I stress the importance of telling the interwoven stories of production, social reproduction, and place-making in communities of origin and destination.

In framing the story of Rustbelt revitalization and its underlying social reproduction processes, my approach is inspired by Gillian Hart's (2006) relational comparison, and by new insights into relational comparisons that Ayse Çaglar & Nina Glick Schiller (2018) offer in their Relational Multiscalar Analysis, which locates migrants and non-migrants within the same analysis of city-making processes. As presented in further detail in my *Global Heartland* (2016), I use a multi-scalar analysis developed through a multi-sited methodology to see not only the node, but also the web; that is, not only the processes that capture and consume immigration and the labour force of migrants, but also those that produce the migrants (see also Guarnizo and Smith 1998; Burawoy et al. 2000; Gille 2001). Multi-scalar analyses and multi-sited ethnographic approaches reveal the indivisible processes of dispossession and displacement that create the contemporary processes of displacement and global labour mobility.

I see the relational approach at the core of radical feminist urban scholarship. An approach that stresses production and social reproduction, as well as local and global processes, practices, and policies cannot explain urban development in isolation. Urban development must instead be treated with recognition of the entanglements of these analytic forces and scales. While anchored in specific locations, my analysis of the global spans local and national boundaries, to recognize the multiple dimensions through which globalization is constituted. For this interscalar analytic move, we need to pay attention to sites of mediation, which are specific institutions and spaces where relationships and entanglements, constituted at various scales, are renegotiated (see contributions to Lamphere 1992). For example, global macro-policies displace people and move them from their communities of origin (displacement), and the micro-politics of localities often allow displaced people to make place in communities of destination across the globe (in-placement). By focusing on spaces and institutions through which these relations of displacement and in-placement are renegotiated, we can move between the local, extra-local, and trans-local, and connect them to one another. In the case of Beardstown, as will be discussed later, these sites of mediation were clearly found in spaces of social reproduction, including the residential arrangements, schools, soccer fields, home-based childcare centres, and public parks and plazas. A relational framework that pays attention to sites of mediation and their specificities of micro-politics allows us to analyse the complex and multiple spatialities and temporalities of globalization (Miraftab 2016). It allows us to account for the nonlinear zigzags that shape global policies and human migration, and to recognize the structures that selectively include and exclude nations in the imagination of a global community.

The relational approach that I advocate here, however, is not merely an intellectual conviction, but also a lived one. As a displaced Iranian who has now made a home in the United States, my positionality cannot be described simply as a privileged Western academic or a displaced woman. In 'Can You Belly Dance? Methodological Questions in the Era of Transnational Feminist Research' (Miraftab 2004), I engage extensively with complexities of doing feminist research in a transnational era where displacements are multiple and power relations in research sites move in multiple directions – some unpredictably so. For example, when doing ethnographic work amongst single mothers in Mexico's poor neighbourhoods, my position in the eyes of my research subjects was not only one of power, as in being an educated woman from a Western university, but also one of powerlessness as a childless woman (at the time) and a woman from Iran, a country that primetime Mexican TV had portrayed as being brutally oppressive of women.[7] In the research that I share here, I similarly inhabit multiple positions vis-à-vis my research subject. I am an insider in that I consider myself a displaced immigrant woman. I am an outsider in that I am an educated professor at a Western academic institution. I am an insider because, like my research subjects, I have made a home in the United States, yet I also have a transnational imagination of home, an 'elsewhere', which is my community of origin. Reflecting on my transnational lived experiences, I cannot understand either myself or my research subject without moving between the worlds that make me who I am today and influence where my imagination lies. In the current era of trans-locally situated imaginations, expectations, hopes, families, and networks, understanding and making visible the full dynamics of social reproduction requires a relational approach that moves across scales of analysis and between points that shape our realities.

Social Reproduction and Feminist Urban Scholarship

Feminist urban scholars have long established that gendered labour and social reproduction are central to production of urban space (see McDowell 1983; Little, Peake, and Richardson 1988; Massey 1994; Kofman 1998; Peake 2009; Peake and Pratt 2017). They have shown how social reproduction is both key to processes of production and accumulation and central to making and unmaking urban spaces. A central condition for (re)production of urban space is, however, making social reproduction invisible through racialized, patriarchally gendered normalization (Kobayashi 2005). Patriarchy's ability to make this central condition invisible is fundamental to the system of capitalism. Under capitalism, the patriarchal gendered systems of sociocultural values and norms make social reproduction work, most commonly performed by women, invisible and natural. Feminist urban scholars point out how patriarchal gendered normalization facilitates the spatial fixes that capitalism uses to address its problems of accumulation. Two examples of such processes are: i) the suburbanization of the United States following World War II; and ii) community-based informal provision of basic urban services for poor communities in the global South following neoliberal structural adjustment policies. The former saw a spatial fix for a post-war crisis of capitalism through spatial restructuring, largely to increase consumption through spatial separation of workplace (in cities) and social reproduction (in suburban homes and neighbourhoods). Scholars of this period of American history have pointed out the role of gender and racial discourse to naturalize and idealize the suburbanization experience that was predominantly aimed at women (Friedan 1963; Hayden 2002; Harris 2013; Manning Thomas 2013). Similarly, Benería and Feldman (1992) later documented the structural adjustment policies implemented in the global South starting in the late 1970s that sought to mend another crisis of capitalism through social expenditure cuts. Such policies transferred the cost of urbanization and service provision onto women in marginalized poor communities, a process that extended their unpaid social reproduction work beyond home, justified as municipal housekeeping (Miraftab 2005) – in other words what Mies had critiqued as *housewifization* extended to municipal services.[8]

The withdrawal and/or redefinition of the state's role in the provisioning of social care (i.e. city and state support for social reproduction) has precipitated a crisis of social reproduction (see Martin 2010; Chant with McIlwaine 2016; Norton and Katz 2017) and has created new geographies of social care (Lawson 2007; England 2010). In both global North and global South, the privatization of state responsibilities for social infrastructures encompassing water, schools, healthcare, waste collection, and more has caused much of the responsibility for collective care to fall back on women, a process Bakker (2003) describes as a re-privatization, whereby the responsibility for social care is transferred from public institutions to the private sector and the domestic sphere of the home and family. For families who cannot afford the private sector, this responsibility shifts to women in the domestic sphere, and for those who can afford privatized services, the responsibility often falls on Black and Brown women who will shoulder these privatized social care jobs. It is the unpaid labour of care that women provide, both to their own families in the domestic realm and to their un-serviced neighbourhoods and towns in the public realm, that makes the social reproduction of low-income populations possible (Miraftab 2004; Mitchell, Marston, and Katz 2004). Feminist urban scholarship, looking at these dynamics of social reproduction as they play out in the processes of urbanization and urban development, has shown the key role played by the unpaid labour of women and shown how, to make this unpaid labour invisible, it has been normalized and naturalized in public discourse.

The question of central concern to this chapter, however, is the way in which gendered and racialized labour invested in social reproduction is made invisible in the contemporary era of global capitalism. This phenomenon has become increasingly complex, often involving transnational processes (Bakker and Silvey 2008). The great mobility of labour and capital – through so-called migration and footloose capital – makes possible processes that fragment not only production but also social reproduction, connecting workers and workplaces in one part of the world to families of workers in another. Increasingly, this means that the care work invested in the social reproduction of workers and their families is manipulated spatially and temporally to be performed from afar – a process that makes global mobility of labour possible but also makes social reproduction work further invisible. How trans-local families contribute to urban and community development can, however, be brought into view by following not only the resources that displaced workers send home (remittances) but also the ways in which communities of origin support displaced workers and hence contribute to the urban development in workers' communities of destination.

The literature on the development-remittance nexus has documented how the employment of a vast body of labour predominantly performing gendered labour of care in one part of the world has, through remittances, led to the development of communities on the other side of the world. The sociological insights gained into the gendered care work by Filipino nannies (Parreñas 2001; Ehrenreich and Hochschild 2003; McKay 2016) or Mexican domestic workers (Hontagneu-Sotelo 2001) have provided a poignant analytic point for understanding the remittance-based development of communities. Here, families in the global North export the burden of their social reproduction work to families in the global South. Calling it the global chain of care work, scholars map socially how transnationally recruited careworkers employed within the domestic spheres or the private institutions of health and care (hospitals and care institutions for elderly, children, or disabled) leave their own children back home to be cared for by their extended families while they care for global North families. The small remittances they send home not only raise their children from afar but also develop housing and basic physical services and structures, referred to as transnational community development (Sandoval 2012).

How these families and their transnational social reproduction practices develop urban spaces and communities in places of destination is, however, less discussed by that same literature. The literature reviewed earlier in the chapter recognizes that displaced workers revitalize Rustbelt towns and that these workers, like all others, rely on the unpaid work of their families for social reproduction. Nevertheless, the invisible resources invested in the social reproduction of workers and their living spaces from afar by their transnational families are seldom acknowledged. This transfer of responsibility from employers in communities of destination to families in communities of origin yet again keeps social reproduction squarely on the shoulders of poor women amongst subordinate populations and keeps it out of sight from the communities of destination.

Such invisibility is also pertinent to the formulation of nativist discourse, which sees the remittances sent home to immigrants' communities of origin as siphoning resources from the United States, but fails to see how those beneficiaries abroad contribute to the communities of destination. Indeed, they naturalize global labour mobility by not considering the dispossession processes that have displaced these workers. Rather, they naturalize the process as one of stress-free mobility across a featureless surface according to the laws of supply and demand: 'people move because there are more jobs in destination X.' The feminist theorization of contemporary social reproduction processes, however, shows how these processes of global labour mobility rest on racialized, gendered, and classed social hierarchies of global capitalism and its social reproduction (Kobayashi 2005; Lebaron and Roberts 2010; Miraftab 2016). I now turn to a specific example to show how the spatial restructuring of social reproduction, when social reproduction is not performed where production takes place, makes social reproduction work and its contribution to production less visible and how these invisible social reproduction processes underlie the ongoing spatial revitalization of urban places in the US Rustbelt.

Outsourced Social Reproduction and Revitalization of Urban Space

Transnationally and trans-locally supported displaced workers and their families are able to revitalize and regenerate communities of origin because they keep their communities of destination alive, economically, socially, and culturally. In the case of Beardstown, depopulated neighbouring towns had turned into ghost towns with their houses boarded up, schools shut down, and only a few shops operating. Beardstown, however, thanks to the arrival of new families, had a healthy housing market; it has not only kept its school open, but built a brand-new school and a library with a vibrant multilingual clientele amongst English-speaking, French-speaking, and Hispanic families. The stores and grocery stores, the gas station, and other businesses were fully functioning as the new population made a home in Beardstown and used its various urban services and spaces. The public square and the public parks held annual cultural festivities and tournaments for newly-established soccer leagues in soccer fields. The urban revitalization and comeback of places like Beardstown needs to be understood in relation to the unpaid social reproduction work performed elsewhere across the globe.

Dispossession and displacement at far-flung sites across the globe are forces that motivate labour to relocate and accept high-risk, low-paying jobs, but these terms do not explain why and how these jobs can be adequately filled. What are the practices and processes that make these wages viable for a foreign-born workforce, but less so for their US-born counterparts? An important invisible aspect of this process is the social reproduction work performed by an army of people and a range of institutions in immigrants' communities of origin. The unpaid work of social reproduction performed and supported in one location (communities of origin) subsidizes wages received across national borders, helping us understand how processes of revitalization rely on care work by displaced workers and their families *in situ* and by families, friends, neighbours, and institutions in remote communities back home. Transnational reorganizations of familial and community care are critical to this particular spatial fix. In Rustbelt communities, such as Beardstown in central Illinois, we see the comeback of a previously depopulating town, where the local workforce left because they saw the wages for the difficult, dirty, and dangerous job of meatpacking as unviable. Then a new, foreign-born workforce was recruited, accepting those very jobs. In this context, the stories of displaced workers that I interviewed from the local plant in Beardstown, and their families in communities of origin in Mexico and Togo, are instructive in understanding why the plant wages were unviable for one group but acceptable for another.

Like 67% of the other West Africans I surveyed in Beardstown in 2008, Margaret and Kwame left part of their family behind – including, at the time, their newborn and their six-year old – when they received their diversity visa.[9] They borrowed money from wherever and whomever they could to make the journey to the United States. They hoped the El Dorado they imagined the United States to be would soon allow them to save money to pay back the debt and send money for their children to join them. But this was far from the reality. Kwame's sister continued raising their children while both worked to pay back debt and pay their older child's school fees. At the time I spoke with them, it had been more than five years since they had made the journey, and they were routinely sending US$50 monthly back home.

Reflecting on my conversations with two US-born workers was enlightening in terms of how plant wages were viable for displaced workers with transnational families, like Margaret and Kwame, and not for the others. The two US-born workers I spoke with included one American of European descent born and raised in Beardstown who had left the plant after wages were cut in the late 1980s, and another was an African-American who had recently relocated to Beardstown from Detroit, but was contemplating leaving the job. Both explained to me that the strategy adopted by families with transnational connections was not a strategy that could work for them. The Detroit worker explained that if she had left her children behind in Detroit, the cost to her would stay the same as in Beardstown. But Margaret, Kwame, and other displaced workers whom I interviewed could not imagine that they could sustain their families in the United States, because raising their children and caring for their elderly parents across the world was a fraction of the cost it would be in the United States. The stark difference in the cost of social reproduction across two wage zones is not new; urban workers have long left rural communities that supported urban wages – for example, Bantustans of the apartheid model relied on this strategy (Legassick and Wolpe 1976). What is new is that the two wage zones are spread across national borders. In short, central to Margaret's and Kwame's ability to accept high-risk, low-paid jobs is the possibility for 'outsourcing' care work to their home communities. Many immigrants leave their children behind in the care of relatives and spouses; others return home to their relatives and families when they are old or injured. Hence, responsibility and cost for such care work rests elsewhere – in other words, displaced labour is made viable through the outsourcing of social reproduction.

The fragmentation of the life cycle across times and spaces is another instance of the global restructuring of social reproduction. Global transnational social reproduction may rely on performing portions of the workers' own life cycle in different locations. For example, Roberto and Javier are two brothers displaced from Michoacán, Mexico, to Beardstown. I met with their family of origin, who explained the processes of land dispossession and economic dispossession through NAFTA that had left them destitute. Neither their milk production nor their agricultural production was competitive in the local market. Hence, Roberto and his brother braved the northern border and found jobs in the Beardstown meat-packing plant. In the case of Javier and Roberto, low-wage US employers were subsidized through the social reproductive work at the beginning and end of Javier and Roberto's life cycles elsewhere. Their births at public institutions and their basic education were performed in Mexico, but once raised as able labourers, they sold their labour in the US market. They would then take their injured, tired, and old bodies to be cared for in their home communities and institutions.

By transnationalizing their social reproduction work, displaced workers subsidize their wages and their ability to remain in backbreaking, low-waged jobs. The possibility of the global restructuring of social reproduction is a key factor in the resurgence of accumulation processes in America's Rustbelt – spaces previously left behind by capital through its restructuring of production.

But social reproduction should not be understood in merely economic terms or through its economic logic. Emotional logics and cultural logics are important aspects of workers' emergent local and trans-local social reproduction strategies and practices. In discussing Roberto and Javier's departure for the United States, their father clearly explained how the economic pressures motivated their displacement. Their mother, however, further explained another aspect of their motivation: she told how young men like Roberto and Javier would feel shame and stigma if they did not brave the borders and cross to the United States. This is just one example of how social reproduction relies on processes and is motivated by incentives that cannot be reduced to economic logic.

Many factors move immigrants to and keep them in places of destination. These range from remittances that they can send home to pay for daily bread, road pavement infrastructure, development, and health and education insurance to the hope that motivates immigrants to embark on risky journeys and perform hazardous jobs out of a sense of obligation. Moreover, feelings of honour and pride can drive a worker to tolerate the disillusionment and meet cultural expectations. Or the imagination of an 'elsewhere', somewhere one can retire or retreat to, is an important consideration in determining the wages workers are willing to work for.

Workers with a transnational imagination often assume a future in the imagined elsewhere where they will be 'set for life' at a lower cost than what is needed in communities of destination. A real or imagined 'future elsewhere' has the material power to make a wage that is untenable for one worker acceptable to another. The imagining of an alternative place that workers create or dream of creating, regardless of its connection to reality, becomes an asset that distinguishes the viability of wages across groups of workers. In short, not only transnational families and their possibility to care for children and elders of displaced workers, but also a real or imagined transnational imagination of a future elsewhere, and cultural expectations and obligations, hope and fear as well as sense of pride and stigma, all play a role in accepting wages that others might not accept. These and other complex rationalities cannot be captured by a cost-based analysis and are key to keeping displaced labourers in low-waged jobs that others might not accept. The globalization and urban revitalization literature has documented how the restructuring of production depopulated and created Rustbelt America, but has paid less attention to these components of the global restructuring of social reproduction, which are crucial not only to the regeneration of depopulating Rustbelt towns but also city centres thriving on low-waged service industries.

Conclusion

The story of the urban revitalization through outsourcing of social reproduction by low-wage displaced workers is not limited to the Rustbelt. Transnational families and globally restructured social reproduction are, in fact, behind contemporary urban development processes, from gentrified neighbourhoods to metropolitan centres in many parts of the world. Many sectors of the global economy, from service and tourism industries to construction, health, and care, thrive on the cheap labour of a displaced workforce with transnational families and imaginations. From Abu Dhabi's soaring tourism-based urbanization, which relies on displaced construction workers whose families are raised elsewhere, to New York City, whose the service industry could not survive without a displaced labour force making beds in its hotels and washing dishes in its restaurants while their families are being cared for back home, urban space is often kept alive and thriving through the work of people across the globe where families of these workers are raised and dreams of their future are kept alive.

Behind these urban vitalities – from the Rustbelt comeback to metropolitan city centres and gentrified inner-city neighbourhoods – are families of the low-waged labourers working not only in these cities, towns, and neighbourhoods, but also across the globe. The work of transnational families is fundamental to processes of production and accumulation as well as urbanization, but the work is made invisible because it is naturalized under such patriarchal norms and values as 'what women do for their family'. These poor families across the globe, with women playing a central role in them, invest their care work, day-in and day-out, into processes that serve as survival strategy through remittances received, but also serve as a subsidy to spatial revitalization of cities across the world where displaced workers sell their labour. The communities across the globe reap rewards from the invisible gendered and racialized social reproduction work performed locally and trans-locally.

Seeing interrelations between production and the social reproduction of the labour force, and interrelation amongst spaces of production and social reproduction, has significant implications for the relational theorization of place and placemaking. Places such as those of immigrants' communities of origin, we are often told, are developed through resources immigrants send home, where it funds roads, infrastructure, houses, and even sports fields. In these stories, as circulated in the US-based mainstream media and popular discourse, immigrants are cast either as villains who siphon resources to their places of origin or as heroes who sacrifice their own leisure and work hard to send remittances home. Neither scenario makes visible how people in communities of origin provide for development of places in immigrants' communities of destination – a perspective that could offer an antidote to the hateful nativist discourse on the rise today.

Moving across vantage points to see connected lives gives us an alternative perspective on resource flows and the range of actors contributing to the revitalization of urban space. We see how the unpaid or underpaid work performed by people in the global South contributes to social reproduction of working people and 'revitalized' places in the global North. Transnational families' practices subsidize immigrants' wages and hence allow them to stay in low-wage jobs and make a home in communities of destination – a process I call *in-placement* in communities of destination. The displaced workers' in-placements and the often-devastating processes of displacement to which they have been subjected should not be seen in isolation from each other. Following dispossession and displacement, these displaced workers invest in houses and public works in their communities of origin, but they also invest in place-making in their destination.

A relational theorization recognizes the multidirectional flows of resources, accounting not only for resource contributions by immigrants (as in remittances) to spatial development of immigrants' communities of origin but also for the gendered social reproduction work that is performed in communities of origin. This work is invisibilized not only because it is gendered and naturalized in a patriarchal system, but also because it is performed in communities that are racialized in the global hierarchies that make flows of resources from global South to North (in)visible. In the spatial development of immigrants' communities of destination, it is the indispensable contribution made by gendered and racialized work performed in communities of origin that too often disappears from view in the global North.

Acknowledgements

I am grateful to the editors of the volume for their careful review of the manuscript and the helpful comments they offered. Responsibility for any shortcomings, however, remains with the author.

Notes

1   Here reference to the Rustbelt as consisting of spaces 'abandoned by capitalism' is not to be misunderstood as spaces outside capitalism. These are rather spaces produced by the restructuring of the global capitalist economy and its attendant different regimes of accumulation.

2   Throughout this text I use the term 'small towns' to refer to 'nonmetropolitan areas'. A nonmetropolitan area, as designated by the Office of Management and Budget, refers to an area outside a standard metropolitan statistical area and that does not contain any settlement whose population exceeds 50 000 individuals (for legal definition of nonmetropolitan areas, see Law Insider n.d.; for census definition of Metropolitan areas, see United States Census Bueareu n.d.).

3   Check cashers serve those paid by cheque but who lack access to banks to cash cheques and write cheques. The most common check casher is Ace Cash Express, a global conglomerate with more than 1300 stores in America. They have operations across four continents fueled by massive capital infusions from bank giants J.P. Morgan Chase and Wells Fargo. Like pawnshops, they appeal to the most disadvantaged populations: those with bad or no credit (including tardy borrowers denied chequeing accounts, the indebted, ex-prisoners). To cash cheques in these stores, fees range typically from 3 to 10% of a cheque's value. Services also extend to writing cheques for customers (typically for a USUS$10.00 or USUS$15.00 fee). This is an enormously popular service; household bills (heating bills, electric bills, mortgage payments, tax payments) are commonly paid this way.

4   The most prevalent pawnshop in the two cities is Cash America, a combination pawnshop and payday lender based in Fort Worth. In 2013, it collected revenues from payday loans totalling US$878 million, compared to only US$21 million in 2003 (Wilson and Miraftab 2015, p. 32). The most

common payday lender in Chicago and Indianapolis, Advance America, posted profits of US$35 million in 2010 (Seeking Alpha 2011). Payday lenders service economically depressed, credit-starved populations by providing short-term cash loans at high interest rates. Payday loans are fast infusions of money (typically US$200 to US$300) often needed for household emergencies (e.g need to pay rent, need to pay electric and gas bills). The loans are to be paid back within two to three weeks, at an interest rate of 15% to 20%. But many borrowers, money-strapped and unable to repay, have to refinance, often at an additional 20%. Refinancing loans ('rollovers') is often repeated. In many instances, debts between US$1200 and US$2500 are incurred on a US$300 loan.

5   Temporary labour agencies and day labour sites anchor this economy as definers of wages and material realities. They deliver inexpensive, warm bodies on a just-in-time basis for manual work, mainly construction, warehouse, assembly-line jobs (Peck and Theodore 2001).The temp agency, Manpower for example, which has a vast global empire operating across the world, had fourth quarter earnings in 2011 of US$5.5 billion, a 5% increase from its previous quarter (Manpower Group 2012).

6   In the contemporary global order of free-market capitalism, complex movements of people across territories – some through voluntary relocation, others through systematic displacement – have continued. However, distinct in their incentives and trajectories, these population movements are commonly referred to as (im)migration – a term that often conflates the varied stories behind people's movements within and across political, social, and cultural borders.

7   See an earlier discussion of my positionality and transnational feminist methodology in Miraftab (2004).

8   Mies's (1980) theory of housewifization grew out of her research on home-based work performed by poor women in the city of Narsapur, India. While they produced lacework for the global market, defined as housewives who use their spare time to earn some extra change, they were paid a fraction of the wage they would otherwise have been paid. Mies refers to these processes as housewifization, whereby the 'ideology of the housewife' supports the exploitative integration of lace workers into global production.

9   Diversity Visa, commonly called Lottery Visa, is a quota system established by the US State Department in 1991 and implemented in 1995 to grant US residency to the underrepresented populations. To afford this emigration, many lottery visa winners leave part of their immediate family behind, typically the youngest of their children, as costs for travel by an entire family are prohibitive. Destinations like Beardstown with a guaranteed year-round employment hold a special attraction for lottery visa winners, as they hope their savings in the United States will fund the emigration of the rest of the family. But with typically limited English and massive debts from the original journey, saving money quickly becomes a distant dream. The acquisition of Diversity Visas involves a labyrinth of trans-local practices and networks that is documented for the Togolese community by the work of anthropologist Charles Piot in Lomé (2019) and mine in Beardstown (2016).

References

Arruzza, C., Bhattacharya, T., and Fraser, N. (2019). *Feminism for the 99%: A Manifesto*. London: Verso.

Bakker, I. (2003). Neo-liberal governance and the reprivatization of social reproduction: Social provisioning and shifting gender orders. In: *Power, Production, and Social Reproduction: Human In/security in the Global Political Economy* (eds. I. Bakker and S. Gill), 66–82. London: Palgrave Macmillan.

Bakker, I. and Silvey, R. (2008). Introduction: Social reproduction and global transformation—from the everyday to the global. In: *Beyond State and Markets: The Challenges of Social Reproduction* (eds. I. Bakker and R. Silvey), 1–16. New York: Routledge.

Benería, L. and Feldman, S. eds. (1992). *Unequal Burden: Economic Crises, Persistent Poverty and Women's Work*. Boulder, CO: Westview Press.

Bhattacharya, T. (2017). Mapping social reproduction theory. In: *Social Reproduction Theory: Remapping Class, Recentering Oppression* (ed. T. Bhattacharya), 1–20. London: Pluto Press.

Burawoy, M., Blum, J., George, S. et al. (2000). *Global Ethnography: Forces, Connections and Imaginations in a Postmodern World*. Berkeley, CA: University of California Press.

Çaglar, A. and Glick Schiller, N. (2018). *Migrants and City-Making: Dispossession, Displacement, and Urban Regeneration*. Durham, NC: Duke University Press.

Castells, M. (1983). *Cities and the Grassroots*. Berkeley, CA: University of California Press.

Chant, S. with McIlwaine, C. (2016). *Cities, Slums and Gender in the Global South: Towards a Feminised Urban Future*. London: Routledge.

Durand, J., Massey, D., and Capoferro, C. (2005). The new geography of Mexican immigration. In: *New Destinations: Mexican Immigration in the United States* (eds. V. Zúñiga and R. Hernández-León), 1–20. New York: Russell Sage Foundation.

Ehrenreich, B. and Hochschild, A.R. (2003). Introduction. In: *Global Woman: Nannies, Maids, and Sex Workers* (eds. B. Ehrenreich and A.R. Hochschild), 1–15. New York: Henry Holt.

England, K. (2010). Home, work and the shifting geographies of care. *Ethics, Place & Environment* 13 (2): 131–150. doi:10.1080/13668791003778826

Federici, S. (2014). The reproduction of labor-power in the global economy, Marxist theory and the unfinished feminist revolution. In: *Workers and Labour in a Globalised Capitalism: Contemporary Themes and Theoretical Issues* (ed. M. Atzeni), 85110. New York: Palgrave Macmillan.

Fennelly, K. and Leitner, H. (2002, December). *How the food processing industry is diversifying rural Minnesota*. (Juliana Samora Research Institute, Working Paper No. 59). https://www.iatp.org/sites/default/files/258_2_102980.pdf (accessed 10 October 2020).

Frey, W.H. (2006). *Diversity Spreads Out: Metropolitan Shifts in the Hispanic, Asian, and Black Populations since 2000*. Washington, DC: Brookings Institute.

Friedan, B. (1963). *Feminine Mystique*. New York: W.W. Norton & Company.

Gille, Z. (2001). Critical ethnography in the time of globalization: Toward a new concept of site. *Cultural Studies – Critical Methodologies* 1 (3): 319–334.

Gozdziak, E.M. and Martin, S.F. eds. (2005). *Beyond the Gateway: Immigrants in a Changing America*. Lanham, MD: Lexington Books.

Grey, M. and Woodrick, A. (2005). Latinos have revitalized our community: Mexican migration and Anglo responses in Marshalltown, Iowa. In: *New Destinations: Mexican Immigration in the United States* (eds. V. Zúñiga and R. Hernández-León), 133–154. New York: Russell Sage Foundation.

Griffith, D. (2008). New midwesterners, new southerners: Immigration experiences in four rural American settings. In: *New Faces in New Places: The Changing Geography of American Immigration* (ed. D. Massey), 179–210. New York: Russell Sage Foundation.

Guarnizo, L. and Smith, M.P. (1998). The locations of transnationalism. In: *Transnationalism from Below* (eds. M.P. Smith and L. Guarnizo), 3–34. New Brunswick, London: Transaction Publishers.

Harris, D. (2013). *Little White Houses: How the Postwar Home Constructed Race in America*. Minneapolis: Minnesota University Press.

Hart, G. (2006). Denaturalizing dispossession: Critical ethnography in the age of resurgent imperialism. *Antipode* 38 (5): 977–1004.

Hayden, D. (2002). *Redesigning the American Dream: Housing, Work, and Family Life*. New York: W.W. Norton & Company.

Herrera, G. (2008). States, work and the social reproduction through the lens of migrant experience. Ecuadorian domestic workers in Madrid. In: *Beyond States and Markets: The Challenges of Social Reproduction* (eds. I. Bakker and R. Silvey), 93–107. London: Routledge.

Hochschild, A.R. (2000). Global care chains and emotional surplus value. In: *On the Edge: Living with Global Capitalism* (eds. W. Hutton and A. Giddens), 130–146. London: Jonathon Cape.

Hontagneu-Sotelo, P. (2001). *Domestica: Immigrant Workers Cleaning and Caring in the Shadow of Affluence*. Berkeley, CA: University of California Press.

Hontagneu-Sotelo, P. and Avila, E. (1997). 'I'm here, but I'm there': The meanings of Latina transnational motherhood. *Gender and Society* 11 (5): 548–571.

Hopkins, C.T. (2015). Introduction: Feminist geographies of social reproduction and race. *Women's Studies International Forum* 48: 135–140.

Irazábal, C. ed. (2014). *Transbordering Latin Americas: Liminal Places, Cultures, and Powers (T)Here*. London: Routledge.

Karger, H.J. (2005). *Shortchanged: Life and Debt in the Fringe Economy*. New York: Berrett-Koehler.

Katz, C. (2001). Vagabond capitalism and the necessity of social reproduction. *Antipode* 33 (4): 709–728.

Kobayashi, A. (2005). Anti-racist feminism in geography: An agenda for social action. In: *A Companion to Feminist Geography* (eds. L. Nelson and J. Seager), 32–40. London: Blackwell Publishing.

Kofman, E. (1998). Whose city? Gender, class and immigration in globalizing European cities. In: *Cities of Difference* (eds. R. Fincher and J.M. Jacobs), 279–300. New York: Guilford.

Lamphere, L. ed. (1992). *Structuring Diversity: Ethnographic Perspectives on the New Immigration.* Chicago, IL: University of Chicago Press.

Law Insider (n.d.). *Legal definition of Nonmetropolitan area.* www.lawinsider.com/dictionary/nonmetropolitan-area (accessed 10 October 2020).

Lawson, V. (2007). Geographies of care and responsibility. *Annals of the Association of American Geographers* 97 (1): 1–11. doi:10.1111/j.1467-8306.2007.00520.x

Lebaron, G. and Roberts, A. (2010). Toward a feminist political economy of capitalism and carcerality. *Signs* 36 (1): 19–44.

Legassick, M. and Wolpe, H. (1976). The Bantustans and capital accumulation in South Africa. *Review of African Political Economy* 3 (7): 87–107. doi:10.1080/03056247608703302

Little, J., Peake, L., and Richardson, P. (1988). Introduction: Gender and the urban environment. In: *Women in Cities: Gender and the Urban Environment* (eds. J. Little, L. Peake and P. Richardson), 1–20. London: Macmillan.

Manning Thomas, J. (2013). *Redevelopment and Race: Planning a Finer City in Postwar Detroit.* Detroit, MI: Wayne State University Press.

Manpower Group (2012). *Annual report.* https://investor.manpowergroup.com/financial-information/annual-reports (accessed 1 September 2015).

Martin, N. (2010). The crisis of social reproduction among migrant workers: Interrogating the role of migrant civil society. *Antipode* 42 (1): 127–151.

Massey, D. (1994). *Space, Place, and Gender.* Minneapolis, MN: University of Minnesota Press.

McDowell, L. (1983). Towards an understanding of the gender division of urban space. *Environment and Planning D: Society and Space* 1 (1): 59–72. doi:10.1068/d010059

McKay, D. (2016). *An Archipelago of Care: Filipino Migrants and Global Networks.* Bloomington, IN: Indiana University Press.

Mies, M. (1980). *Lace Makers of Narsapur: Indian Housewives Produce for the World Market.* London: Zed Books.

Mies, M. (2007). Patriarchy and accumulation on a world scale revisited. (Keynote lecture at the Green Economics Institute, Reading, 29 October 2005). *International Journal of Green Economics* 1 (3/4): 268–275.

Miraftab, F. (1996). (Re)Production at home: Reconceptualizing home and family. *Journal of Family Issues* 15 (3): 467–489.

Miraftab, F. (2004). Can you belly dance? Methodological questions in the era of transnational feminist research. *Gender, Place and Culture: Journal of Feminist Geography* 11 (4): 595–604.

Miraftab, F. (2005). Making neoliberal governance: The disempowering work of empowerment. *International Planning Studies* 9 (4): 239–259.

Miraftab, F. (2012). Small-town transnationalism: Socio-spatial dynamics of immigration to the heartland. In: *The Transnationalism and Urbanism* (eds. S. Kratke, K. Wildner, and S. Lanz), 220–231. London: Routledge.

Miraftab, F. (2016). *Global Heartland: Displaced Labor, Transnational Lives and Local Placemaking.* Bloomington, IN: Indiana University Press.

Mitchell, K., Marston, S., and Katz, C. (2004). *Life's Work: Geographies of Social Reproduction.* Oxford: Blackwell.

Norton, J. and Katz, C. (2017). Social reproduction. In: *International Encyclopedia of Geography* (eds. D. Richardson, N. Castree, M.F. Goodchild et al.). London: Wiley. doi:10.1002/9781118786352.wbieg1107

Parreñas, R.S. (2001). *Servants of Globalization: Women, Migration, and Domestic Work.* Palo Alto, CA: Stanford University Press.

Peake, L. (2009). Urban geography: Gender in the city. In: *The International Encyclopedia of Human Geography* (eds. R. Kitchin and N. Thrift), 320–327. London: Elsevier.

Peake, L. and Pratt, G. (2017). Why women in cities matter. In: *Urbanization in a Global Context* (eds. A. Bain and L. Peake), 276–294. Toronto: Oxford University Press.

Peck, J. and Theodore, N. (2001). Mobilizing policy: Models, methods, and mutations. *Geoforum* 41 (2): 169–174.

Piot, C. (2019). *The Fixer: Visa Lottery Chronicles.* Durham, NC: Duke University Press.

Portes, A. and Rumbaut, R. (2006). *Immigrant America: A Portrait.* Berkeley, CA: University of California Press.

Pratt, G. (2005). From migrant to immigrant: Domestic workers settle in Vancouver, Canada. In: *A Companion to Feminist Geography* (eds. L. Nelson and J. Seager), 123–317. Oxford, UK: Blackwell Publishing.

Prugl, E. (1999). *The Global Construction of Gender: Home-Based Work in the Political Economy of the 20th Century.* New York: Columbia University Press.

Ruggles, S., Alexander, T., Genadek, K. et al. (2010). *Integrated public use microdata series: Version 5.0.* [Machine-readable database]. Minneapolis, MN: University of Minnesota. IPUMS-USA, University of Minnesota. www.ipums.org.

Sandoval, G.F. (2012). Shadow transnationalism: Cross-border networks and planning challenges of transnational unauthorized immigrant communities. *Journal of Planning Education and Research* 33 (2): 176–193.

Sandoval, G.F. and Maldonado, M.M. (2012). Latino urbanism revisited: Placemaking in new gateways and the urban-rural interface. *Journal of Urbanism: International Research on Placemaking and Urban Sustainability* 5 (2–3): 193–218.

Sassen, S. (2014). *Expulsions: Brutality and Complexity in the Global Economy.* Cambridge, MA: Harvard University Press/Belknap.

Seeking Alpha 2011. *Payday lender Advance America: Numbers look good, but can you stomach the business?* (6 June). http://seekingalpha.com/article/273539-payday-lender-advance-america-numbers-look-good-but-canyou-stomach-the-business (accessed 1 September 2015).

Stull, D., Broadway, M., and Griffith, D. eds. (1995). *Any Way You Cut It: Meat Processing and Small-Town America.* Lawrence, KS: University Press of Kansas.

United States Census Bureau (2020). *Metropolitan and micropolitan.* www. census.gov/programs-surveys/metro-micro.html (accessed 10 October 2020).
Wilson, D. and Miraftab, F. (2015). New inequalities in America's Rust Belt. In: *Cities and Inequalities in a Global and Neoliberal World* (eds. F. Miraftab, D. Wilson, and K. Salo), 28–48. London: Routledge.

7

From the Kampung to the Courtroom

A Feminist Intersectional Analysis of the Human Right to Water as a Tool for Poor Women's Urban Praxis in Jakarta

Meera Karunananthan (Blue Planet Project)

Introduction

In October 2017, the Supreme Court of Indonesia ruled to terminate private water concessions in Jakarta, arguing that privatization had led to violations of the human right to water (RTW). The ruling was celebrated by human rights and water justice groups around the world (see Harsono 2017; TNI 2017). However, in a highly unusual development, the Supreme Court reversed its own decision less than a year later through a judicial review filed by the Ministry of Finance. This chapter highlights the involvement of Solidaritas Perempuan, an Indonesian feminist organization, in its invocation of the human right to water to demand greater access to and control over the urban water system for poor women in order to ease their heavy social reproduction burden. Although a final ruling overturned the October 2017 ruling, appearing to confirm the views of critics that human rights are ineffective when it comes to challenging the privatization of water and sanitation services, I argue that by using a human rights framework the organization was able to foreground poor women's everyday experiences, thus exposing the gendered violence of capitalism at the urban, household, and bodily scales.

A Feminist Urban Theory for our Time: Rethinking Social Reproduction and the Urban, First Edition. Edited by Linda Peake, Elsa Koleth, Gökbörü Sarp Tanyildiz, Rajyashree N. Reddy & darren patrick/dp.

Through a feminist place-based investigation, this chapter argues that the use of socio-economic rights by poor women in Jakarta serves to shift analytical focus and rescale knowledge production, to visibilize the intensification of poor women's burden of social reproduction, and to demonstrate how socio-economic rights may compliment feminist right to the city projects by asserting the use value of city services for marginalized communities and as part of a feminist transitional programme towards a just city. In so doing, this chapter highlights the importance of social reproduction as an arena for struggles against urban capitalism in cities of the global South. It calls for Marxist debates regarding revolutionary praxis to be re-examined in light of both the constraints faced by women living in the margins of cities of the global South as well as their aspirations.

In the following section, I set out my methodology and clarify my positionality as a researcher and my relationship to the Jakarta-based study discussed in this chapter. This is followed by a theoretical discussion informed by the works of feminist scholars, outlining gaps in dominant critical urban studies that have contributed to invisibilizing the urban politics of poor women. I then turn to the privatization of water and anti-privatization struggles in Indonesia and highlight the work of Solidaritas Perempuan Jakarta (SPJ), a national feminist network advocating for poor women's rights to water in Jakarta. Finally, I glean key lessons regarding feminist urban praxis from Solidaritas Perempuan's prominent campaign for the right to water in Jakarta. The story of Solidaritas Perempuan's engagement with human rights instruments provides insights into the use of a human rights framework by an urban poor women's movement, to articulate and organize around their daily embodied experiences with privatized urban drinking water systems in the arena of social reproduction, by foregrounding sites and scales overlooked in traditional right to the city and anti-privatization discourses.

Methodology and Positionality

I have been working at the intersection of water justice advocacy and research for more than a decade as an employee of the Council of Canadians (COC), a Canadian social justice organization. Since 2011, I have been running the Blue Planet Project (BPP), a global water justice project supported by the COC, which played a leading role in the campaign to have the human rights to water and sanitation formally recognized in international law (Barlow 2013). In my role at the BPP, I have supported grassroots struggles for water justice through community-relevant research, by mobilizing international solidarity and helping to amplify frontline voices through international media (see, for example, CBC News 2012; Yglesias 2014; Karunananthan 2017).

I have been collaborating with groups in Indonesia since 2011, when the BPP supported the national coalition People's Forum for the Right to Water (KruHa) to document their struggle against a World Bank imposed water law. Solidaritas Perempuan works closely with KruHa on water issues. I met members of Solidaritas Perempuan at a meeting convened by KruHa in April 2015 in Jogjakarta. When I returned to Indonesia in December 2015 to work with KruHa, I interviewed members of Solidaritas Perempuan and its Jakarta branch, Solidaritas Perempuan Jakarta, regarding their community-based research project. In 2017, the BPP translated the organization's research into English. Anecdotally, I was aware through my work with anti-poverty organizations and community activist groups in Detroit and Cape Town that poor racialized women were leading struggles against the privatization of water. Yet, there was far too little acknowledgement of their work in social movement or academic literature.

Analysis presented in this chapter draws from data from Solidaritas Perempuan's research and from interviews with staff Aliza Yuliana, Dinda Nuuranissa Yura, and Arieska Kurniawaty. I am also grateful for the support of colleagues Muhammad Reza Sahib and Sigit Budiono at KruHa Indonesia who assisted this research by sharing relevant documentation, providing analysis, and facilitating interviews.

Water, the Urban, and Social Reproduction

Within critical urban scholarship, various biases have resulted in the failure to acknowledge poor women from the global South as political agents in struggles against capitalism. For decades, feminist scholars have emphasized that locating women's revolutionary politics requires a shift in analytical focus to places, spaces, and scales ignored in dominant political economic literature (Nagar et al. 2002). Feminist urban scholars have highlighted parallel gaps in masculinist theorizations of the urban under capitalism (see Chant 2013; Peake and Reiker 2013; Peake 2016). In addition, I argue in this section that the failure to centralize social reproduction stems from a long history within Marxist thought of disregarding sites of subaltern revolutionary praxis, which are also invariably gendered and racialized. Racialized women of colour are therefore invisibilized in dominant critical urban scholarship at multiple levels. Given these scholarly oversights, I revisit the debate regarding the potential of rights-based strategies for achieving urban transformation, arguing that this literature would benefit from a feminist intersectional approach that centres social reproduction as a key site of anti-capitalist struggle.

There has been a recent surge in academic literature and social movement activism inspired by Henri Lefebvre's (1996 [1967]) concept of the *right to the city*. In the 1960s, Lefebvre was concerned with the ways in which urban space was subordinated to the interests of capital. He viewed the city as a site of multiple conflicts and contradictions. On the one hand, the city is a centre of commerce; on the other hand, it is a site of artistic expression and community life, which contradict efforts to organize urban space for profit maximization. In Marxist terms, the city contains 'use value' for those who inhabit it, which contradicts initiatives to extract profits by maximizing 'exchange value'. To Lefebvre, these sites of conflict are rife with opportunities to disrupt the capitalist urban agenda. Notwithstanding concerns raised by some critical scholars regarding the ways in which the right to the city concept has been depoliticized or neutralized as a result of its varied application by a range of actors (see Purcell 2002; Walsh 2013), a large body of contemporary Marxist scholarship emphasizes its promising potential as a framework for radical urban politics.

By conceptualizing revolution outside the industrial factory, Lefebvre's work diverged from the Marxist traditions preceding him (Purcell 2002; Harvey 2012; Marcuse 2012). Subsequent Marxist scholars, such as David Harvey (2012), for whom contemporary capitalism rests heavily on urbanization processes, also view the right to the city as an appropriate tool for revolutionary struggle: 'Since the urban process is a major channel of use,' Harvey argues, 'then the right to the city is constituted by establishing democratic control over the deployment of surpluses through urbanization' (Harvey 2012, p. 23). He describes a shift in focus from industrial workers ('who build the pipes') to workers who bring running water to the home without acknowledging the unpaid work of women who draw and carry water to the homes without piped water. Although Harvey offers a meek nod to social reproduction, it appears largely as an auxiliary to struggles taking place within the sphere of production.

While seeing greater potential for discussion of gendered urban subjectivities in this shift from the factory as the heart of revolutionary struggle, feminist scholars argue that Lefebvre and the extensive scholarship inspired by his theories have largely overlooked the urban struggles of women and other marginalized communities (Buckley and Strauss 2016). Urban feminist scholars echo concerns raised by Nagar et al. (2002), who argue that the vast scholarship on economic globalization ignores women's experiences and contributions by excluding the places, spaces, and scales where women's work took place. They argue that the concept of the 'urban' is constructed around spaces traditionally dominated by men such as the industrial core or public streets, whereas feminized spaces such as the home are overlooked or coded as non-urban and therefore outside the purview of urban studies (Peake and Reiker 2013; Buckley and Strauss 2016; Peake 2016). In addition, focusing on productive labour within formal sectors, the literature has overlooked the contributions of women in the informal sector. These oversights have limited the discussion of social reproduction, its relationship to capitalist production, and the implications for the formation of gendered subjectivities within and through urbanization processes.

Social reproduction refers to the daily and long-term labour that goes into reproducing the work force. It comprises multiple practices of largely gendered care work – the often unpaid labour that goes into sustaining people, homes, and communities (Katz 2001; Di Chiro 2008; Roberts 2008). It includes the work that goes into meeting the food, shelter, and healthcare needs that are vital to the survival of workers, as well as the cultural education and socialization that enables the labour force to perform its expected functions. Feminist scholars have emphasized that shifts in the social reproduction functions of the neoliberal state have forced shifts in women's labour. When essential services are priced beyond the reach of poor households, poor women are forced to work harder and longer in order to ensure the survival of their families. Neoliberal processes that restrict access to services that are vital to survival such as drinking water provision, are based on 'an assumption of women's infinite and elastic labour supply' (Miraftab 2010, p. 4).

The failure to centre social reproduction is also consistent with a long history of Marxist scepticism towards the urban subaltern classes – the so-called lumpen proletariat, whom Marx and Engels perceived as counter-revolutionary and famously referred to as the 'dangerous classes' and 'social scum' whom they deemed to be far too concerned with survival to lead or even support a proletariat revolution. As the Manifesto of the Communist Party puts it, this class 'may here and there be swept into the movement by a proletariat revolution; its conditions of life, however, prepare it far more for the part of a bribed tool of reactionary intrigue' (Marx and Engels 2019 [1850], p. 71).

While not nearly as hostile as Marx himself, many leading urban scholars continue to offer theoretical justifications for overlooking the leadership of communities living in the margins of the city. Although Bayat (2000, 2017) does not view subaltern communities as apolitical, he is nonetheless sceptical of their potential to engage in formal politics outside certain exceptional circumstances found in Latin America. In African and Asian cities, Bayat (2000, p. 545) sees a more passive, informal, and atomized form of activism he labels 'quiet encroachment', defined as 'the silent, protracted but pervasive advancement of the ordinary people on the propertied and powerful in order to survive and improve their lives.'

Similarly, Harvey and Wachsmuth (2012) argue that informal sectors including domestic services and 'back street sweatshops' are vital to urban struggle, yet at the same time are less inclined to engage in sustained collective organizing efforts and less likely to develop class solidarities by virtue of their precarious and individualized conditions. Likewise, Peter Marcuse's (2012) classification of urban struggle distinguishes those who are 'deprived' or 'excluded' from the benefits of urban life from those who are 'alienated', only the latter being more inclined towards revolution.

As Tilley (2017) argues, the deprivation of urban subaltern subjects is deeply gendered. The common thread in these dominant theories of subaltern urban politics is the failure to see leadership or revolutionary potential emerging out of sites of social reproduction and informal labour, predominantly occupied by women. Harvey (2014, p. 197) for instance, states 'social reproduction is unlikely to be a source of revolutionary sentiment,' explaining that Lefebvre's emphasis on everyday lives point to social reproduction only as a site from which to build a critique of capitalism.

The place-based study of women's struggles against water privatization in Jakarta discussed in the following pages shows otherwise. Not only does it demonstrate the capacity and willingness of those facing the greatest deprivation to fight back against global capital, it demonstrates the significance of social reproduction as an important arena for organizing against capitalism at multiple scales.

This study also offers insights into a vigorous debate regarding the value of human rights within anti-capitalist struggles. Broadly, critiques range from those who view human rights as outrightly harmful (see De Souza 2008; Kneen 2009) to those who see them as ineffective (Bakker 2007b; Bond 2014). With regards to the RTW, a middle ground is found in the works of Morinville and Rodina (2013), Mirosa and Harris (2012), Angel and Loftus (2019), and others, who argue that the RTW has been useful in contextually specific ways. Radha D'Souza (2008, p. 8) charges that human rights do not change 'the class character of the states [...] and regimes of expropriation [...] so fundamentally that the left can abandon its skepticism of human rights in political activism.' Likewise, legal scholar Samuel Moyn argues human rights have historically replaced languages of 'distributional equality' with a minimalist language of 'distributional sufficiency'. They provide 'a bare floor of material protection', he maintains, but are inadequate in placing a 'ceiling' on the accumulation of wealth (Moyn 2018). He explains, the focus on 'securing enough for everyone' lacks ambition and offers moral cover to states promoting market fundamentalism while recognizing human rights (Moyn 2018, p. xii). Karen Bakker (2007b), however, argues that the global campaign for the RTW is an incoherent response by anti-globalization activists to the threat of the privatization of water. Noting that human rights are compatible with private property rights and private sector participation in the supply of drinking water, she argues that RTW activists suffer from 'conceptual confusion' (2007b, p. 433). The crux of her argument is that it is the commons, not the RTW, that is the real antidote to privatization. Echoing Bakker, Patrick Bond (2014) points to the disappointing results of long drawn out legal challenges invoking the right to water by South African anti-metering activists.

However, much like the scholarship championing the right to the city, this body of work largely lacks the gendered intersectional approach that is critical in understanding the value of socio-economic rights within struggles for urban justice. Specifically, when it comes to questions of uneven access to water, an emerging body of research demonstrates not only that more feminist research is needed but that any research is meaningless unless it is feminist, intersectional, and engaged with social reproduction. Farhana Sultana, Mohanty, and Miraglia's (2013) place-based research regarding the experiences of women living in the Korail slum of Dhaka reveals the need for discussions of alternatives to water privatization to take into account the 'gendered possibilities and constraints' that shape women's everyday relationships to water governance. Along the same vein, Faranak Miraftab (2010) demonstrates the specificities of women's collective action in cities of the global South. She argues that the deep spatial and economic informality in these cities, combined with shifts in production-reproduction relations in urban development have opened up new spaces for women's active citizenship and collective action. Putri (2020) describes how community activists in Jakarta's kampungs have engaged in what Miraftab (2009) refers to as 'insurgent planning' to intervene strategically within formal and otherwise hostile processes. Unlike 'quiet encroachment', Miraftab (2009, p. 44) describes insurgent planning as a transgressive counter-hegemonic practice which seeks to 'disrupt domineering relationships of oppressors to the oppressed, and to destabilize the status quo through consciousness of the past and imagination of an alternative future.'

Elson and Gideon (2004, p. 15) observe that socio-economic rights have had a prominent place in feminist movement organizing for several decades. They argue that far from being liberal feminists seeking equal access to property rights, many feminist groups use such claims 'to challenge the Operation of Contemporary Capitalism. They focus on rights to use collective property such as public health systems, rather than on rights to own private property.' In short, feminist scholars have shown that women organize in complex and context-specific ways, navigating multiple constraints and contradictions in order to challenge capitalism in meaningful ways. Rather than dismissing these contradictions as signs of 'conceptual confusion', place-based feminist intersectional research serves to better understand the forces that shape women's ability to produce the city. For example, to Moyn's (2018) charge that human rights offer a minimalist programme, Trotsky's (1938) distinction between minimum and transitional is useful. A minimum programme is one that seeks minor reforms without challenging the capitalist system, while a transitional programme is one that mobilizes against the bourgeois regime and serves as a bridge towards a more just society. Drawing from SPJ's work, I argue that being denied the basics of subsistence in the city and forced to work harder and longer, women living in the margins need and seek out transitional strategies. Hence, an approach based on socio-economic rights, demanding public services that ease the burden of social reproduction, is more aptly described as one of transition than a minimum. Pointing to the corporate appropriation of human rights discourse, D'Souza (2018) suggests the left should ask 'why do they [corporations] want human rights?' I flip this question in the next section to shift the focus from corporations to poor racialized women in Jakarta. Why do *they* want human rights?

The Privatization of Water and Anti-privatization Struggles in Indonesia

Swyngedouw (2005, p. 82) describes privatization as a

> process through which activities, resources and the like, which had not been formally privately owned, managed or organized, are taken away from whomever or whatever owned them before and transferred into a new property configuration that is based on some form of private ownership or control.

The privatization of urban water and sanitation systems takes various forms of either direct private ownership or the outsourcing of operations to private entities. The latter is referred to as private-public-partnerships, or P3s, which typically involve long-term (20- to 30-year) concessions to private firms for one or several functions including the financing, design, building, and operation of water and/or wastewater services. Miraftab (2010, p. 6) further emphasizes that privatization entails both contracting out to private corporations and outsourcing the 'private sphere of women's free work'. Hence privatization is more than an enclosure of the commons; it is an appropriation of poor women's labour. It sets up a parasitic relationship through which private corporations extract profits abetted by the capitalist state, exacting a toll on women's bodies through work that is naturalized by patriarchal norms within homes and communities.

Urban water systems became vulnerable to privatization strategies in the last two decades of the 20th century as investors began to seek new channels for capital surplus absorption and the realization of surplus value (Swyngedouw 2005). In Indonesia and elsewhere in the global South, international financial institutions catalyzed these processes through structural adjustment programmes, which made loans contingent upon economic liberalization measures, including the privatization of public infrastructure. In the late 1990s, the Suharto government began negotiations with foreign transnationals making grand promises to drastically expand and improve water and sanitation services in Jakarta (Heriyanto 2018; Marwa and Tobin 2018). At the time, only 42% of people living in the city had access to piped water (United Nations Development Program [UNDP] 2008). In 1997, without an open bidding process, an agreement was signed between PAM JAYA, the state utility, and two multinationals, the French company Suez Lyonnaise des Eaux and the British Thames Waters. During the 2-year negotiation process, World Bank-sponsored international consultants acted as 'advisors' to the Government of Indonesia. In order to circumvent Indonesian legislation preventing foreign corporations from operating in the country, the corporations partnered with local businesses close to Suharto. Suez signed an agreement with the Salim group owned by a close friend of Suharto's and became PT PAM Lyonnaise Jaya (Palyja). Thames Waters, which was later sold to the consortium PT Aetra Air Jakarta, partnered with Suharto's son whose company had no prior experience in the sector. PAM JAYA, the public company was to serve as a regulatory body (Harsono 2003). In order to ensure full-cost recovery and a 22% return on investments, an unusual arrangement was made with the private companies. To this day, the public operator pays private concessionaires a fixed water charge per unit delivered regardless of revenues from tariffs, which are approved and collected by the city (UNDP 2008). In addition, water charges are indexed to protect them from inflation, creating an insurmountable public deficit, which has disincentivized expansion to poor residents who pay lower tariffs according to the law (Furlong and Kooy 2017).

Although privatization was brought in with promises of universal access that justified extremely high rates of return on investments, data from PAM JAYA shows less than 60% of the city's population has access to water today, with network expansion largely favouring wealthier neighbourhoods. Bakker et al. (2008, p. 14) describe the Jakarta water supply system as an 'archipelago rather than a homogenous network', noting the large pockets of the city without access to services. In addition, 'access', as defined by the presence of piped water, does not account for consumption patterns which are determined by a range of other factors including affordability and availability in sufficient quantities. Densely populated informal settlements, referred to as kampungs, are scattered throughout the city, sprawling out between the city's main roads, formally planned and wealthy areas. Of its population of 10 million inhabitants, it is estimated at least a quarter of Jakarta's residents live in these informal settlements. Kampung residents continue to rely on a variety of alternative sources, including public hydrants connected to the formal network, drinking water kiosks, and untreated water from canals for laundry and cleaning (Putri 2017). When asked about the decision to focus their work on privatization, a coordinator of the SPJ explained that private contracts prevented the negotiation of an urban water system that would better serve the needs of people in kampungs in contemporary Jakarta (Yuliana, private communication, 20 May 2017). Indeed, private contracts justified by promises of universal coverage locked in patterns of uneven access instead.[1]

Solidaritas Perempuan Jakarta and Poor Women's Rights to Water

Solidaritas Perempuan is a national feminist network representing 5700 grassroots women including peasants, fisher-folk, migrant workers, and poor urban women. The organization has conducted political education workshops for women since 2004. These workshops are designed to enable working-class women to identify and to engage in political issues that impact upon their lives. Frontline women determine priorities for political action through these workshops, relying on the organization's staff to provide technical support and secure resources to translate priorities into political strategies.

According to national coordinator, Dinda Nuuranissa Yura (personal communication, 27 December 2018), access to drinking water emerged as a prominent theme through workshops held in Jakarta. Women with access to piped water often described being forced to stay up late at night until they could access water due to irregular and unreliable flow patterns in their neighbourhoods; they frequently shared stories regarding health concerns such as skin rashes, attributed to the quality of drinking water; women who were buying from vendors to supplement their needs were paying for water at exorbitant rates forcing them to work longer hours in precarious jobs; and the daily struggle to access sufficient water to meet social reproduction needs was placing a significant psychological and physical burden on women. These factors were identified by the Jakarta branch of the organization, SPJ, as a key priority for the political engagement of their members. As a result, SPJ has been actively involved in a variety of strategies including legal action demanding universal access to public water services.

Legal Challenges Against Privatization

Indonesian social movements have been engaged in two parallel legal processes relating to water in the last decade: one dealing with the constitutionality of Water Resource Law No. 7, a law imposed by the World Bank in 2004, and the other challenging private water concessions in Jakarta.

Both cases consisted of citizen lawsuits, which enable a private citizen or group of citizens to take legal action against violations of the public interest (Sundari 2018). Unlike a class action lawsuit which is filed on behalf of a particular group of individuals, a citizen lawsuit is filed on behalf of all citizens of a state seeking the enforcement of human rights or environmental legislation rather than financial compensation for damages (Sundari 2018; Pinakunary 2020). Although civil society groups have had success in achieving significant policy changes through this tactic since 2003, Sundari (2018) argues that as a relatively new practice in Indonesia borrowed from other legal traditions, there are inconsistencies in its application. Indeed, the legal community appears to be divided on key elements. While some legal experts argue that a citizen lawsuit can be launched against private entities (Triayu and Indrawan 2019), others argue that the private sector cannot be the subject of a citizen lawsuit (Pinakunary 2020). Sundari (2018) calls for a legislative provision to strengthen the consistency and predictability of citizen lawsuits in Indonesian law. In the meantime, both water policy related citizen lawsuits sought to reverse the legacy of World Bank structural adjustment programmes in Indonesia.

In 2004, the Indonesian government adopted Water Resource Law No. 7, deregulating water resource management at the behest of the World Bank as a condition for a US$300 million loan. In response, the People's Forum for the Right to Water (KruHa) – a national water network of which Solidaritas Perempuan is a founding member – in collaboration with the Indonesian Forum for the Environment (WALHI) and the Jakarta Water Consumers Coalition organized to have the law annulled on the premise that it would violate constitutional provisions protecting public rights to water.

Indonesia's 1945 constitution was ahead of its time in its recognition of socio-economic rights (KruHa and Blue Planet Project 2012). Prior to the 1948 UN Declaration on Human Rights, the Indonesian Constitution established the duties of the State to ensure basic needs and to govern natural resources in the public interest. Notably, Article 33 states that 'all vital sources of production essential to the country and lives of people shall be controlled by the state' and that 'land, water and natural resources shall be controlled by the state for the greatest benefit of people.' This historical legal context is important in understanding why social movements in

Indonesia engaged in legal battles to roll back neoliberalization through human rights claims.

In 2005, the Constitutional Court stopped short of annulling the law but offered a detailed review affirming the human right to water of Indonesian citizens and the obligation of the state to treat water as a common good under Article 33 of the constitution (KruHa Indonesia and Blue Planet Project 2012). Given that the law had not yet been implemented, the Court concluded it was 'conditionally constitutional' provided its implementing regulations upheld Article 33 and other constitutional provisions. The 2005 Constitutional Court ruling paved the way for a more decisive victory in 2015, when civil society groups returned to court to challenge Water Resource Law No. 7 for failing to meet the conditions of constitutionality established in the previous ruling. The court concurred and ruled to annul Water Resource Law No. 7. Amongst other provisions, the Court also argued that a 2005 government regulation, which sanctioned private sector participation on the basis of full cost recovery, contravened constitutional law.

The second legal action was launched in 2013 by Koalisi Masyarakat Menolak Swastanisasi Air Jakarta (KMMSAJ), a city-wide citizen's coalition of which SPJ is a member. The citizen lawsuit involved twelve residents including two members of SPJ. The lawsuit sought the annulment of private concessions for drinking water supply based on testimonies of kampung residents, including members of SPJ (Heriyanto 2018). The Supreme Court ruled that private water companies had violated the human right to water by failing to improve access, creating conditions for excessive water tariffs including a US$90 million loss for the city-owned public sector partner through unreasonable contractual obligations (Johnson 2015). This was seen as a major victory until the Supreme Court unexpectedly reversed its decision in a judicial review requested by the Indonesian Ministry of Finance due to technical errors in the proceedings (IDN Financials 2019). According to the judicial review, the case failed to meet the criteria for a citizen lawsuit due to the addition of private water corporations as defendants. As previously noted, Indonesian legal experts are divided on this matter. In common law jurisdictions, from which Indonesia has borrowed the practice, notably the United States, citizen lawsuits do enable legal action against private corporations (Sundari 2018). In addition, Yenny Silvia Sirait, a spokesperson for LBH, the law firm representing plaintiffs in the case, argues that the review differs from the position of the Jakarta District Court in its previous ruling on the matter (personal communication 29 September 2020).

Despite this final ruling, the government of Jakarta has declared that it would maintain its commitment to initiate a process of returning water into public hands. While it is not yet clear what these plans entail, this declaration appears to be the result of the tremendous public pressure generated by the high-profile legal challenge. Furthermore, Muhammed Reza Sahib, the coordinator of KruHa Indonesia, argues that the judicial review did not invalidate the Supreme Court's ruling that privatization of water had resulted in violations of the human right to water (personal communication, 15 December 2018).

Community-based Research on the Impacts of Privatization

In 2015, SPJ requested that the national secretariat of Solidaritas Perempuan help develop a survey that would enable them to document and report on violations of the RTW in five kampungs in which their members were active: Kebon Jeruk, Koja, Tebet, Cilincing, and Penjaringan. SPJ believed the findings of this survey could inform their political work and enable them to gather evidence for legal proceedings. In response, the national secretariat sought technical assistance from Elva Community Engagement, a European online platform supporting community mapping projects. With Elva's support, Solidaritas Perempuan created an online survey tool that could be accessed with mobile phones. A questionnaire was developed based on the normative content of the RTW,[2] which helped women record their daily experiences with the availability, quality, and affordability of their primary drinking water source. Solidaritas Perempuan Jakarta members were trained by the organization's staff to facilitate town hall meetings through which they gathered qualitative input, provided access to mobile phones, and assisted women using the online survey. Between September 2015 and January 2016, 1158 women participated in the survey-based study. Information was also gathered through the townhall discussions.

Within a short period of time, Solidaritas Perempuan Jakarta collected information from more than 1000 women. It is important to note that this research was driven by an organization run by women who are amongst the surveyed. It represents what Mohanty (2003) refers to as reading from the bottom up, which calls for the writing and sharing of testimonials of everyday struggles to visibilize those who are erased from hegemonic struggle.

Analysis of the data gathered, in addition to identifying significant barriers faced by women seeking to access drinking water, showed how human rights norms, such as affordability, quality, and accountability, interacted with the neoliberalization of the urban water system and linked to the intensification of their burden of social reproduction. Table 7.1 summarizes key observations and experiences reported by participants.

Many surveyed women accessed the formal network through public standpipes or third-party connections. Some had private household connections. Through a perverse pricing system, standpipes used by poor people are priced at much higher rates than household connections, making water far more expensive by the unit for poor households. Yet, as Bakker et al. (2008) note, the overall cost of securing a network connection, as well as the precarious tenure of many kampung residents, make direct connections unviable. Women with running water in their homes complained about infrequent services of unreliable quality and very low pressure requiring them to wait for long periods of time for buckets to fill. In addition, a large number of connections in poor neighbourhoods are illegal. Hence, even though many women described networked water as their primary source, their access to water is often precarious.

In two of the sub-districts surveyed, the primary source of drinking water was a private utility, in two others the state utility, PDAM, was the primary source of drinking water and in the remaining one, participants relied primarily on well water. Overall, the data showed no difference in women's perceptions and experiences with private and public utilities.

However, as Aliza Yuliana, an SPJ staffer shepherding the research project explains, the terms and conditions of access to water have been shaped by the neoliberalization of the system as a whole (personal communication, 20 May 2017). Hence the public utility does not operate independently from its private counterpart. It is part of a system that is organized around commitments to private investors.

Drinking water quality is generally poor throughout Indonesia; 55% of drinking water samples in Indonesia were found to be contaminated with fecal coliform (ADB 2016). Wastewater management has only recently received modest attention from the state seeking to address exceptionally high rates of groundwater and surface water contamination (van der Wulp et al. 2016). Drinking water from the network is not potable anywhere in the city, but it is meant to be safe when boiled prior to consumption. Solidaritas Perempuen Jakarta (SPJ 2017) notes in addition to the added labour, that the cost of making water safe to consume is a significant financial constraint for poorer households.

Untreated surface water and well water sources, which were a primary source for kampung residents in the 1960s, are no longer safe for consumption (Putri 2017). While wealthier residents tend to opt out of the network using efficient pumps that penetrate cleaner sources of groundwater at deeper levels, poor residents rely on shallow wells that are either becoming more and more contaminated due to inadequate wastewater treatment or becoming saline due to excessive groundwater pumping (Furlong and Kooy 2017). Nonetheless, surveyed well water users reported

Table 7.1 Observations of kampung residents regarding human right to water norms

Normative criteria	Observations	Experiences
Availability	Unpredictable timing of tap water provision Water not consistently available in sufficient quantities and acceptable quality Low water pressure In Tebet, over 50% of households outside service area are forced to rely on groundwater from wells.	Women were forced to supplement main water supply using alternative sources, including rainwater, to meet their daily needs, lengthening the amount of time spent collecting water. Alternative sources were more likely to be unsafe for consumption. A long time was spent waiting for buckets to fill up. Women were often forced to wake up in the middle of the night to collect water.
Quality	The majority do not trust the quality of networked water, even when easily accessible. Many report inconsistency of water quality and frequent encounters with turbidity and foul-smelling water. Taps are frequently clogged, yielding water containing sediments. In one district serviced by PT Aetra women reported frequent worm infestations in the tap water. Well water users were most likely to perceive their water to be safe	Participants reported health concerns due to poor water quality. Having to care for sick family members added to women's burden. Having to boil water added to labour, time, and electricity costs.

Table 7.1 (Continued)

Normative criteria	Observations	Experiences
Affordability	Women felt they paid far too much for water that was not of adequate quality.	Women reported spending longer hours working in informal, precarious jobs to manage shortfalls in household finances to meet basic needs, including those of drinking water.
	Network connection fees were considered prohibitive even, if the end result was cheaper water.	
	Boiling water adds to the cost of drinking water.	
Accountability	Nearly a quarter of participants were not aware of the ability to register complaints with the utility.	Women reported feeling frustrated and disempowered in attempts to engage with the utilities.
	More than half had tried to contact the utilities to register concerns but felt they were not taken seriously.	

higher levels of confidence in the quality and safety of their drinking water than those accessing water through the formal network.

According to SPJ, affordability was reported as a major concern. Putri (2017) notes that kampung users spend 7 to 10% of household income on water, which is at least double the international standard for affordability. Above 3 to 5%, experts warn that users may sacrifice their needs or limit their access to other essentials (Winkler 2012). Although women buying water from third-party sources pay much higher rates than those with access to household taps, as noted above, steep connection fees and precarity of tenure make household connections unviable for many (Kooy and Bakker 2008). The unaffordability or unavailability of water in sufficient quantities forces poor women in Jakarta to use multiple sources of water, including groundwater, rainwater, and surface water that is unsafe for consumption, for cleaning, laundry, and other non-consumptive purposes (Putri 2017). This significantly extends the amount of time spent in activities related to water collection. Town hall discussions revealed that poor women spend on average 4 to 6 hours per day in activities related to collecting water. This includes time spent seeking out alternatives to supplement inadequate or unaffordable water from primary sources.

In addition, when water rates rise, women have to work longer hours to manage budget shortfalls through informal work including food preparation, laundry, or childcare in wealthier homes. Hence, the neoliberal city intensifies the exploitation of poor women engaged in the social reproduction of the formal and informal city through both unpaid and low-wage work. They are required to work long hours to collect water for their own families and long hours in low-paid care work in middle-class homes in order to pay for water and other basic necessities.

SPJ notes that the tremendous time burden of water-related activities prevents women from engaging in their communities and in activities of leisure. Not surprisingly, women revealed a great deal of psychological stress caused by the daily challenges of meeting their basic needs in relation to water.

Finally, international financial institutions including the Asian Development Bank often mandate utilities to offer a complaints mechanism purportedly to enhance public accountability. Although a quarter of the participants were unaware of the service, more than half had made at least one attempt to register a complaint and felt dismissed or treated rudely by staff on the other end. They reported being frustrated by their inability to engage the utility in order to improve their service. These experiences with the utility contribute to a general sense of disempowerment amongst kampung women vis-à-vis the water services that are so critical to their daily lives.

Conclusion

Solidaritas Perempuan Jakarta's research does not offer a generalized conclusion regarding the right to water, but rather a historicized and place-based understanding of its use within the feminist praxis of a Jakarta-based organization. Their rights-based campaign for water offers a lens through which to highlight the embodied experiences of privatization and a tool through which women in marginalized housing conditions could make claims to the city. It should be noted that Solidaritas Perempuan did not set out to establish individual entitlements to clean water, but rather a collective remedy in the form of robust public water and sanitation services that would meet the needs of all. As such, their work is more evocative of a Lefebvrian right to the city campaign than it is a liberal human rights programme. It seeks to challenge the predatory extraction of exchange value from the urban drinking water system by using the human right to water as a tool to assert the use value of those who are most excluded.

Furthermore, the process asserts poor women as human rights experts and rightful authors of human rights discourse, defining norms through their everyday experiences, as called for by feminist scholars (see Nagar et al. 2002; Peake 2016). Solidaritas Perempuan Jakarta's work entailed both a shift in analytical focus to scales, places, and spaces that visibilize poor women's experiences and their legitimate inclusion in knowledge production that is otherwise reserved for human rights lawyers and academics. I draw three observations from a feminist intersectional perspective.

First, SPJ used human rights to visibilize the intensification of women's burden of social reproduction. Their work provides insights into the use of a human rights framework by an urban poor women's movement to build knowledge and organize around their daily embodied experiences with privatized urban drinking water systems. The study linked macro-economic processes with the finer scales, places, and spaces where they are experienced by marginalized women. Human rights language provided useful parameters for community-based knowledge production, investigating and documenting the gendered and class-based violence of Jakarta's neoliberal drinking water system as embodied by women and manifested in everyday experiences within the home and community. This process of visibilizing poor women and positioning them as legitimate experts allowed women to take stock of the price that they were paying on a daily basis to keep their families alive within a privatized system. It drew from their everyday experiences to inform a broader coalition's campaign for large-scale systemic change centred around demanding the same quality of service and security of access to public drinking water for women living in precarious housing conditions as others using this system. In doing so, Solidaritas Perempuan Jakarta provides what Mohanty (2003) refers to as a 'bottom-up' reading of global processes. The process demonstrates how collective action can be driven by the experiences of those most dispossessed.

Second, human rights were invoked to maximize the disruption of exchange value by legitimizing the social reproduction demands of marginalized communities. Kampung women engaged in activities of social reproduction fight a daily and intensifying battle against a state that has historically devalued their lives and criminalized the communities they reproduce (see Tilley 2017). In this context, it is a radical act for poor women to claim equal rights to city services and collectively expose transnational capital as an obstacle to the state's ability to fulfill its human rights obligations.

A May 2018, a *New Internationalist* article cites a Solidaritas Perempuan Jakarta organizer comparing conditions in her neighbourhood to that of residents in a wealthy district: 'we are living in the capital city, why should we have to suffer from such problems?' (Firdaus 2018). Contrary to views that human rights serve to fragment and depoliticize Lefebvre's radical vision for urban revolution (see Purcell 2002), in Jakarta the RTW was mobilized not only as a legal strategy but as a discursive tool in media and political campaigns to assert the legitimacy of subaltern women to shape the city according to their needs and demands. In this case, the right to water served as a channel for marginalized women living in poverty to claim their right to exist within the city, to assert the use value of water over exchange value, and demand the collectivization of social reproduction in the form of public water and sanitation services that meet their daily needs.

Third, Solidaritas Perempuan Jakarta's work shows how socio-economic rights might serve as part of a transitional programme for a just city. The campaign for the right to water in Jakarta was part of a multi-pronged strategy that involved coalition-building, public mobilization, media outreach, and lobbying. In and of themselves, the legal challenges produced mixed outcomes. However, for more than a decade, the legal challenges put poor women exposing predatory private sector practices in the spotlight, invigorating public debate and exerting political pressure on decision-makers. Amongst other outcomes, as of March 2019, Solidaritas Perempuan Jakarta and KruHa have expressed cautious optimism vis-à-vis efforts to re-establish a public water service in Jakarta. Solidaritas Perempuan Jakarta's work shows poor marginalized women as strategic actors claiming their rightful place in the city. Though perhaps not engaged in a full Lefebvrian project of producing the city in their own image, their grassroots campaign is better understood as a transitional strategy (Trotsky 1938) than a minimalist one. A feminist reframing of this concept requires the expansion of Trotksy's focus on the workplace as a site of collision between the bourgeoisie and the proletariat to sites of social reproduction. To Trotsky (1938), worker's rights served a transitional purpose by pitting questions of life and death against the interests of capital, which eventually proved to be 'unrealizable' within the capitalist system, thus leading to its demise. Similarly, Solidaritas Perempuan Jakarta's work demonstrates how the RTW can serve to mobilize against capitalism by exposing its drive to devalue the lives of subaltern populations. For the worker in the realm of social reproduction, who is not unionized and has no access to workers' rights, this campaign demonstrates how socio-economic rights might constitute a viable transitional programme, serving not only to claim use value over exchange value of urban water systems but also to reclaim the labour power of women whose unpaid work has served to subsidize a parasitic system.

Acknowledgements

I am deeply grateful to colleagues at Solidaritas Perempuan and KruHa Indonesia for their support with this chapter: Aliza Yuliana, Dinda Nu-uranissa Yura, Arieska Kurniawaty, Muhammad Reza Sahib, and Sigit

Budiono. I would also like to thank Linda Peake, Elsa Koleth, Rajyashree Reddy, and Luisa Veronis for their insightful feedback.

Notes

1 Inequities and exclusion however are not solely the result of privatization, but of a multi-faceted and complex history of colonial and post-colonial violence against the urban poor (Furlong and Kooy 2017). Prior to the Suharto era, nationalist policies of the more left-leaning Sukarno government consolidated patterns of unequal access to water established by the colonial regimes preceding it. As part of his grand nationalist agenda, Sukarno funnelled public funds to large monumentalist projects which developed the city's modern areas inhabited by the elite, while neglecting the expanding informal areas of the city deemed contradictory to the image of a modern Jakarta.
2 The normative content defines the parameters of the human right to water and serves as the criteria through which a state's compliance with human rights obligations is assessed. (See United Nations Social and Economic Council 2003).

References

Angel, J. and Loftus, A. (2019). With-against-and-beyond the human right to water. *Geoforum* 98: 206–213.

Asian Development Bank (ADB) (2016). *Indonesia Country Water Assessment.* www.adb.org/sites/default/files/institutional-document/183339/ino-water-assessment.pdf (accessed 12 April 2019).

Bakker, K. (2007b). The commons versus the commodity: Alter-globalization, anti-privatization and the human right to water in the global South. *Antipode* 39 (3): 430–455.

Bakker, K, Kooy, M., Shofiani N., and Martijn E. (2008). Governance failure: Rethinking the institional dimensions of urban water supply to poor households. *World Development* 36 (10): 1890–1915.

Barlow, M. (2013). *Blue Future: Protecting Water for People and the Planet Forever.* New York: The New Press.

Bayat, A. (2000). From 'dangerous classes' to 'quiet rebels': Politics of the urban subaltern in the global South. *International Sociology* 15 (3): 533–557.

Bayat, A. (2017). *Revolution without Revolutionaries: Making Sense of the Arab Spring.* Stanford, CA: Stanford University Press.

Buckley, M. and Strauss, K. (2016). With, against and beyond Lefebvre: Planetary urbanization and epistemic plurality. *Environment and Planning D: Society and Space* 34 (4): 617–636.

CBC News (2012). *Rights groups investigate Canadian-owned mine in Mexico.* (25 November). www.cbc.ca/news/canada/rights-group-investigates-canadian-owned-mine-in-mexico-1.1223278 (accessed 12 April 2019).

Chant, S. (2013). Cities through a 'gender lens': A golden 'urban age' for women in the global South? *Environment and Urbanization* 25 (1): 9–25.

Di Chiro, G. (2008). Living environmentalisms: Coalition, politics, social reproduction and environmental justice. *Environmental Politics* 17 (2): 276–298.

D'Souza, R. (2008). Liberal theory, human rights and water-justice: Back to square one? *Law, Social Justice & Global Development Journal.* University of Westminster School of Law Research Paper No. 09-10. www.ssrn.com/abstract=1348610.

D'Souza, R. (2018). *What's Wrong with Rights: Social Movements, Law and Liberal Imaginations.* London: Pluto Press.

Elson, D. and Gideon, J. (2004). Organising for women's economic and social rights: How useful is the International Covenant on Economic, Social and Cultural Rights? *Journal of Interdisciplinary Gender Studies: JIGS* 8 (1/2): 133–152.

Firdaus, F. (2018). An end to Jakarta's water woes? *New Internationalist* (21 May). www.newint.org/features/2018/05/21/ending-jakartas-water-woes (accessed 4 September 2019).

Furlong, K. and Kooy, M. (2017). Worlding water supply: Thinking beyond the network in Jakarta. *International Journal of Urban and Regional Research* 41 (6): 888–903.

Harsono, A. (2003). *Water and politics in the fall of Suharto* (10 February). www.andreasharsono.net/2003/02/water-and-politics-in-fall-of-suharto. html (accessed 24 May 2019).

Harsono, A. (2017). Indonesia's Supreme Court upholds water rights. *Human Rights Watch* (12 October). www.hrw.org/news/2017/10/12/indonesias-supreme-court-upholds-water-rights (accessed 4 September 2019).

Harvey, D. (2012). *Rebel Cities. From the Right to the City to the Urban Revolution.* Brooklyn, NY: Verso Books.

Harvey, D. (2014). *Seventeen Contradictions and the End of Capitalism.* Oxford: Oxford University Press.

Harvey, D. and Wachsmuth, D. (2012). What is to be done? And who the hell is going to do it? In: *Cities For People, Not For Profit* (eds N. Brenner, P. Marcuse, and M. Mayer), 264–270. London: Routledge.

Heriyanto, D. (2018). What you need to know about Jakarta's water privatization. *The Jakarta Post* (12 April). www.thejakartapost.com/news/2018/04/12/what-you-need-to-know-about-jakartas-water-privatization.html (accessed 24 May 2019).

IDN Financials (2019). *Supreme Courts grants petitions of Ministry of Finance regarding water privatization.* (1 February). www.idnfinancials. com/n/22492/Supreme-Court-grants-petition-of-Ministry-of-Finance-regarding-water-privatisation (accessed 24 May 2019).

Johnson, C. (2015). Indonesia: Jakarta court bans water privatization. *Global Legal Monitor* (March 30). www.loc.gov/law/foreign-news/article/indonesia-jakarta-court-bans-water-privatization (accessed 15 October 2020).

Karunananthan, M. (2017). A group of Canadians teachers could decide the future of Chile's water supply. *The Guardian* (12 June). www.theguardian.

com/global-development-professionals-network/2017/jun/12/chile-water-privatisation-canada-teachers (accessed 4 September 2019).

Katz, C. (2001). Vagabond capitalism and the necessity of social reproduction. *Antipode* 33 (4): 709–778.

Kneen, B. (2009). *The Tyranny of Rights*. Ottawa: The Ram's Horn.

Kooy, M. and Bakker, K. (2008). Splintered networks: The colonial and contemporary waters of Jakarta. *Geoforum* 39: 1843–1858.

KruHa Indonesia and Blue Planet Project (2012). *Our right to water: An exposé on foreign pressure to derail the human right to water in Indonesia*. www.canadians.org/sites/default/files/publications/RTW-Indonesia-1.pdf (accessed 24 May 2019).

Lefebvre, H. (1996 [1967]). Right to the city. In: *Writings on Cities* (trans. E. Kofman and E. Lebas), 63–181. Oxford: Blackwell Publishers.

Marcuse, P. (2012). Whose right(s) to what city? In: *Cities For People, Not For Profit* (eds N. Brenner, P. Marcuse, and M. Mayer), 24–41. London: Routledge.

Marwa and Tobin, D. (2018). Jakarta's plan to get more public power in water sector might not work well. *The Conversation* (8 February). www.theconversation.com/jakartas-plan-to-get-more-public-power-in-water-sector-might-not-work-well-89320 (accessed 4 September 2019).

Marx, K. and Engels, F. (2019 [1850]). The Manifesto of the Communist Party. In: *Karl Marx, The Political Writings* (ed. D. Fernbach), 56–91. London: Verso Books.

Miraftab, F. (2009). Insurgent planning: Situating radical planning in the Global South. *Planning Theory* 8 (1): 32–50.

Miraftab, F. (2010). Contradictions in the gender-poverty nexus: Reflections on the privatization of social reproduction and urban informality in South African townships. In: *The International Handbook of Gender and Poverty: Concepts, Research and Policy* (ed. S. Chant), 644–648. Cheltenham, UK: Edward Elgar Publishers.

Mirosa, O. and Harris, L. (2012). Human right to water: Contemporary challenges and contours of a global debate. *Antipode* 44 (3): 932–949.

Mohanty, C. (2003). *Feminism without Borders: Decolonizing Theory, Practicing Solidarity*. New Delhi: Zubaan.

Morinville, C. and Rodina, L. (2013). Rethinking the human right to water: Water access and dispossession in Botswana's Central Kalahari Game Reserve. *Geoforum* 49: 150–159.

Moyn, S. (2018). *Not Enough: Human Rights in an Unequal World*. Boston, MA: Harvard University Press.

Nagar, R., Lawson, V., McDowell, L. et al. (2002). Locating globalization: Feminist (re)readings of the subjects and spaces of globalization. *Economic Geography* 78 (3): 257–284.

Peake, L. (2016). The twenty-first century quest for feminism and the global urban. *International Journal of Urban and Regional Research* 40 (1): 219–227.

Peake, L. and Rieker, M. eds. (2013). *Rethinking Feminist Interventions into the Urban*. London: Routledge.

Pinakunary, F. (2020). Understanding actio popularis or citizen lawsuit. *Fredrik J. Pinakunary Law Offices* (6 March). www.fjp-law.com/understanding-actio-popularis-or-citizen-lawsuit (accessed 25 September 2020).

Purcell, M. (2002). Excavating Lefebvre: The right to the city and its urban politics of the inhabitant. *GeoJournal* 58: 99–108.

Putri, P.W. (2017). Spatial practices and institutionalization of water sanitation practices in Southern Metropolises: The case of Jakarta and its Kampung Kojan. *International Journal of Urban and Regional Research* 46 (6): 926–945.

Putri, P.W. (2020). Insurgent planner: Transgressing the technocratic state of postcolonial Jakarta. *Urban Studies* 57 (9): 1845–1865.

Roberts, A. (2008). Privatizing social reproduction: The primitive accumulation of water in the era of neoliberalism. *Antipode* 40 (4): 535–560.

Solidaritas Perempuan Jakarta (2017). *Monitoring Report: The Right to Water of Women in Jakarta.*

Sultana, F., Mohanty, C.T., and Miraglia, S. (2013). Gender justice and public water for all: Insights from Dhaka, Bangladesh. *Municipal Services Project.* Occasional Paper No. 18.

Sundari, E. (2018). The Indonesian model of the citizen lawsuit: Learning how to adopt and how to adapt. *International Trade and Business Law Review* 21: 323–332.

Swyngedow, E. (2005). Dispossessing H_2O: The contest terrain of water privatization. *Capitalism, Nature, Socialism* 16 (1): 81–98.

Tilley, L. (2017). Immanent politics in the kampungs: Gendering, performing and papping the Jakarta economic subject. In: *Women, Urbanization and Sustainability: Practices of Survival, Adaptation and Resistance* (ed. A. Lacey), 43–65. London: Palgrave MacMillan.

Transnational Institute (TNI) (2017). *Jakarta Supreme Court terminates water privatization.* (17 October). www.tni.org/es/node/23724 (accessed 10 October 2020).

Triayu, R. and Indrawan, I. (2019) *Citizen lawsuit in Indonesia.* (30 August). www.idsattorneys.com/citizen-lawsuit-in-indonesia (accessed 10 October 2020).

Trotsky, L. (1938). The death agony of capitalism and the tasks of the Fourth International: The mobilization of the masses around transitional demands to prepare the conquest of power: 'The transitional program'. *Bulletin of the Opposition* May–June 1938 Edition. *Trotksy Internet Archive.* www.marxists.org/archive/trotsky/1938/tp/transprogram.pdf (accessed 24 May 2019).

United Nations Development Program (2008). *Examples of successful public-private partnerships* 15. *Special Unit for South-South Cooperation.* United Nations Development Program.

United Nations Economic and Social Council (2003). *Substantive issues arising in the implementation of the International Covenant on Economic, Social and Cultural Rights.* General Comment No. 15: The right to water (arts 11 and 12 of the Internationl Covenant on Economic, Social and Cultural Rights). *EC.12/2002/11* (20 January). www2.ohchr.org/english/issues/water/docs/CESCR_GC_15.pdf (accessed 19 October 2020).

van der Wulp, S.A., Dsikowitsky, L., Jurgen Hesse, K. et al. (2016). Master Plan Jakarta, Indonesia: The giant seawall and the need for structural treatment of municipal waste water. *Marine Pollution Bulletin* 110 (2): 686–693.

Walsh, S. (2013). 'We won't move': The suburbs take back the center in urban Johannesburg. *CITY* 17 (3): 400–408.

Winkler, I.T. (2012). *The Human Right to Water: Significance, Legal Status and Implications for Water Allocation.* Portland, OR: Hart Publishing.

Yglesias, M. (2014). Why Detroit is cutting off tap water to thousands of people. *Vox* (8 July). www.vox.com/2014/7/8/5878713/why-detroit-is-cutting-off-tap-water-to-thousands-of-people (accessed 4 September 2019).

8

Re-imagine Urban Antispaces! for a Decolonial Social Reproduction

Natasha Aruri

(K LAB, Institute for City and Regional Planning, TU Berlin)

Introduction: Linking the 'Anti-Politics Machine' and Socio-Spacio-Cide

Although in itself not a new phenomenon, today there are increasing numbers of people on the move due to environmental, economic, and political crises – not the least of which are droughts, floods, land and water contamination, imperial (proxy) warfare, and settler (neo)colonialism. It is estimated that an average of 25 million persons are displaced annually due to disasters, that is 1 of every 300 persons each year (UNISDR 2017). Displaced persons seek safety and sustenance mostly in less affected urban areas. As Arjun Appadurai expounds (2003, 2008), cities are mostly 'translocalities', constituting geographies of arrival and departure. While in Palestine there is no documentation on displacements due to natural disasters, the Israeli colonial apparatus (a system of inflicting differentially scaled human-made crises over time) employs displacement as a tool to socio-geographically re-engineer the Palestinian presence and ensure economic exploitation and subjugation. Amongst its normative practices are: the demolition of material structures and communities within Area C[1] by the Israeli army;

A Feminist Urban Theory for our Time: Rethinking Social Reproduction and the Urban,
First Edition. Edited by Linda Peake, Elsa Koleth, Gökbörü Sarp Tanyildiz, Rajyashree N. Reddy & darren patrick/dp.

the confiscation of Palestinian agricultural lands for the expansion of Israeli colonies in the West Bank or under the pretext of creating 'security zones'; the demolition of properties under the pretext of their owners' lack of licences or through the creation of legal and administrative hurdles that prevent Palestinians from obtaining construction permits; and arrests (with and without charges, of adults and minors), assassinations, checkpoints, curfews, bans on travels, and the complex permits system (for work, treatment, visits, imports, and exports, etc.), to name a few. While many communities have been stagnating, shrinking, or erased, Ramallah has been expanding and has become an 'enclave micropolis' to which many 'have hitched their fortunes' (Taraki 2008, p. 14), expanding in the space separating it from the surrounding Israeli colonies that aim at erasing its description as 'Jerusalem North'. In Palestine, polycrises constitute the everyday rhythm of life; cityscapes are saturated with colonial social relations across all scales; and othering and socio-spatial disconnects have become the channels through which economic exploitation, the furthering of dependencies, and the Israeli colonial project are enabled.

While there is considerable research on how urban marginality is systemically and socially reproduced, there is less exploration of how it is spatially reproduced. In light of recent wars and advancements in technologies of destruction, some urban scholars have been investigating concepts such as 'urbicide' (Graham 2006) and 'spacio-cide' (Hanafi 2009). Such studies tend to focus on aspects of spatial annihilation, its impact on economies and the subjugation of urban morphologies to the logics of surveillance and policing (e.g. Graham 2006), and the way spatial perception impacts narratives of identities and accentuation of traumas (e.g. Abourahme 2011). In Ramallah, there is little investigation of how neoliberalism – which plays out as new imperial capitalism – is spatializing its domination through the urban, causing socio-cide (Hilal and El-Sakka 2015); where residents are socially disconnected, contestation increased, and opportunities to challenge the status quo are hindered. Building on feminist critiques of capitalist social formation, this chapter explores how spacio-cide affects the everyday lives of people in Ramallah, and reveals and challenges the features of urban spaces that incubate regressive patterns of social reproduction, spatially engender systemic discrimination, marginality and othering, and inhibit the formation of more just ecosystems.

Today with growing demands to re-adapt cities to be able to tackle climate change and to better utilize resources (through looping, upcycling, etc.), there is an opening to explore alternatives to the spatializations of domination produced under colonialism and neoliberalism, and the forms in which socio-spatial reproduction take place. In this chapter, I focus on how spacio-cide in the city of Ramallah operates through a system of urban spatial management that pays little attention to public space provision, and uses forms of architecture that alienate people and inhibit social interaction and solidarity-building against the neocolonial system. Thereupon, I discuss concepts that aim at inspiring new social contracts for creating city spaces such that they can give rise to more just social reproduction, i.e. social reproduction based on understanding urban life as one of connections, of well-being as an accumulative construct, and of solidarities as key for unravelling new imaginations of the urban.

Following Bakker and Gill (2003), I understand social reproduction as a broad concept that includes the modes, processes, and structures of interaction in everyday life as well as the social spaces that are created for these interactions. These spaces are simultaneously constructed by the system while also constituting a site of resistance to the system. As Bakker and Gill (2003, pp. 17–18, emphasis added) put it:

> Social reproduction [...] refers to both biological reproduction of the species (and indeed its ecological framework) and ongoing reproduction of the commodity labor power. In addition, social reproduction involves institutions, *processes and social relations associated with the creation and maintenance of communities* – and upon which, ultimately, all production and exchange rests.

Given social reproduction's role in capitalist systems of commodity and value production, class formation and wealth extraction, feminist approaches examine relations that go beyond biological reproduction and labourer's subsistence needs to include intellectual, emotional, and physical needs and desires, and bring to the forefront the workplace-household-community continuum of non-monetized and monetized activities (Ferguson and McNally 2014). That said, social reproduction is mediated by power relations that are not only behaviourial, political, and economic, but equally spatial. Therefore, when discussing social reproduction in the urban context, there is a need to consider how the materiality of cityscapes contributes to subjectivity formation, the alienation of labour and (both waged and unwaged) labourers from wealth and securities, and access to possibilities and opportunity. This chapter starts from this expansive definition of social reproduction, which highlights processes and social relations, patterns of exploitation and control, and formation of identities and resistance, focusing in on ways in which this broad understanding dovetails with commodity production and spatial production. In doing so I engage with the counter-topography of Ramallah, which is a method that Katz (2001, p. 1231) argues 'takes for granted that space is both the bearer and reinforcer of social relations, and that if these relations are to be changed so too must their material grounds.'

The chapter develops over four sections. In order to interrupt the systems reproducing spacio-cide, I first address the need to break the 'anti-politics machine', that is the developmentalist doctrine that focuses on the technical and suspends the socio-political (including the colonial reality), and which dominates Palestinian urban development. I then discuss the contextual issues of land management and the militarization of urban space (spacio-cide through infrastructures of surveillance and policing and the stifling of spaces of opposition), and I explore the links between the forms of produced unbuilt urban space and the burdening of (not only but particularly) women's everyday lives. In the third section, I focus on the need to interrupt neocolonial and neoliberal city making by planning urbanisms for uncertainty rather than stability. I discuss the need to understand the urban as a constantly layering space and explore how the current neocolonial and neoliberal hegemony in Ramallah can be challenged through engaging communities in the production of space. I conclude by proposing a re-introduction of the existing *Masha'* laws (that currently apply to land ownership as commons to serve agriculture) in an urban context and to put social reproduction at the centre of city- and place-making processes, and therewith decolonization. By doing so I call for a re-thinking of the dimensions through which social reproduction operates alongside and within commodity and spatial production, by taking the physicality of urban space as the structuring point, the medium of materialization, and for (re-) establishing relationalities and connections against dispossession and spacio-cide.

This chapter builds on multidisciplinary perspectives as I am a Palestinian architect by training, with a focus on urban planning and design, housing, and emergency response. I employ the analysis and findings of my doctoral dissertation titled '100 × Ramallah: imaginations, otherness, and (de)colonization in antispaces of sumud, 1914–2014', which focused on how successive politico-economic regimes changed social norms in my hometown Ramallah, reshaping the spatial features and the nature of spaces in the city.[2] Since this dissertation, I have been researching and exploring particular ways to apply concepts of spatial decolonization and re-imagining alternative urbanization patterns (see Aruri 2015b).

The 'Anti-Politics Machine' in Palestine

After two and a half decades and over US$ 30 billion in Official Development Assistance (ODA) for a population of about 4 million (approximately US$7500 per capita), equality, dignified living conditions, and political stability are farther from reach than they were before the Oslo Accords of 1993. Indeed, what the 'Peace Process' and developmentalist state-building project brought about can be summarized in three points. First, is the creation of what Samara (2000) calls the 'Circus Economy' through the application of the Washington Consensus model in Palestinian communities; where biased economic policies were put into place through the Paris Economic Protocol (Oslo II) on 29 April 1994, creating 'an enduring legacy' by which 'development came to mean that Palestinians became closely monitored guests in their own economy' (Lagerquist 2003, p. 8). This broke up long-term social contracts of collective struggle for justice and self-determination[3] in favour of momentary privatized gains; in which political and economic elites have been consumed with taking as much public wealth as possible while younger generations have been deprived of securities and hooked into debt (what Thatcher called 'popular capitalism' [Aruri 2015a]). Second, as Alaa Tartir has documented and explains, the type of Palestinian institutions and style of governance the international community has been creating and nourishing through the annual ODA has not obtained security for citizens, but has forged 'the paradigm of securitizing everything: a securitized peace process, a securitized development process, securitized spaces, [...] etc. so everything is securitized, but security is absent' (Tartir 2017a). And third, Israel's settler colonial enterprise was legitimized through creating the non-sovereign, dependent, authoritarian Palestinian Authority (PA) and downloading the responsibility for welfare and well-being of Palestinians on it, and Israel was relieved from the monetary costs for sustaining the status quo by exporting these to the international community ODA budgets.

In this constellation, Palestine is a microcosm of happenings around the world: extractive capitalists secure 'free rides' at the expense of native populations that are pushed into exacerbated vulnerability; spaces and communities of resistance to these paradigms are silenced under the pretext of 'security' and being 'alien' to the national 'sui generis'; and affluent governments finance and ensure the continuation of business as usual. While all these issues are at the heart of political orders, the international community (especially the World Bank, IMF, and UN bodies) propagates developmental discourses and governance hierarchies that assume economic progress could be achieved regardless of (non)achievements at the local and national political levels, which is what Ferguson (2009) coins the 'anti-politics machine'.

Under the logics of the 'anti-politics machine', Palestinian local government bodies such as the Ramallah Municipality see their work as technical, possible in spite of geopolitical constraints, and focus on building the bases for a prosperous and internationally connected Palestinian state.[4] They reference indicators such as statistical monetary measurements (e.g. GDP) while ignoring human indicators such as quality of healthcare, education, judiciary systems, and freedom of speech. Given the international focus on Israel's security, a third of the Palestinian Authority's budget is allocated to the (non-productive, hegemonic) security sector (Tartir 2017b), contributing to the systemic under-investments in other domains that are vital for social reproduction.

Furthermore, under the anti-politics machine, gender equality is limited to what Abdo calls 'imperialist feminism' (2014, p. 58), which she considers to be a continuation of orientalist feminism where the focus of interventions lies 'on women's bodies and sexuality through related themes such as the "veil", "harem", "polygamy", and other culturally-based gendered symbols' (Abdo 2014, p. 57). Meanwhile, issues such as economic roles, class, race, and resistance to colonialism are ignored or sidelined under the assumption of the lack of agency of Palestinian women. Development discourses have reduced Palestinian women to recipients of modernity rather than makers of an anti-colonial modernity (as was the image and the case in pre-Oslo times). In effect, during the Oslo era, women's representation in and making of politics has regressed, levels of so-called 'honor crimes' (what local feminists refer to as femicide, e.g. by Prof. Nadera Shalhoub-Kevorkian) have increased (Al-Haq 2015), and with the rise of neo-religiosity in Palestinian politics, issues of custody over children, and equality in inheritance amongst other everyday issues have not improved.

Moreover, foreign-aid-led development discourses continue to disregard the intensifying critique by Palestinians of the fragmentation, privatization, and NGO-ization of the provision of basic services, a provision which is fundamental for environments of just social reproduction. As Bakker and Gill (2003, p. 19) indicate, contemporary development discourses constitute modes of primitive accumulation through several dimensions, the main two of which are privatization and alienation or enclosure of common social property that effectively 'grant more power to capital, while simultaneously undermining socialized forms of collective provisioning and human security.' In Ramallah, socialized frameworks of social reproduction that in pre-Oslo times operated voluntarily and clandestinely.[5] since the Oslo Accords these have been systemically NGO-ized, institutionalized, and are now heavily dependent on ODA for operation (the circus economy). As for spatial alienation, this has been unfolding through mediocre financial investments in social infrastructures, increased spatial militarization, and the systemic creation of antispaces, as explored in the following section.

Achieving new modes of social reproduction is interlinked with combating vulnerability and disenfranchisement. This depends on many factors, the foremost of which is breaking the anti-politics machine through the re-politicization and feminization of local governance and city-making. In spite of globalization, the 'local' is the level at which resources – whether those of 'nature', capital, or labour – are managed, and access or alienation forged. Here professions such as urban planning, design, and architecture continue to convince generations of practitioners that their work is apolitical, philanthropic, and serves the good of the people – equally regardless of gender. Yet historically, these professions have been proximate to regimes (oppressive and otherwise), that continue to play key roles in serving the needs of elites through spatially legitimizing and enabling extraction, and that still mostly operate within patriarchal working environments. Through their work, urban designers and planners have been able to systematically entrench hierarchies and socio-political orders, including gender inequalities, suspend decision-making from collective realms, and monopolize it in the hands of elites (including themselves), and find new regulatory interpretations that engender spatializations of domination.

Socio-cide: Spatial Militarization and Antispaces

In Palestinian localities, technocratic logics (which dominate the work of bodies such as municipalities and NGOs) sideline critiques of apolitical development discourses and prevent consideration of social reproduction in systems relevant for city-making, as demonstrated in attitudes towards managing land ownership, use, and design. In this section, I focus on these logics in relation to the architectural and morphological elements of Ramallah to explain how spatial militarization, the absence of semi-public and transitional spaces, and the lack of accounting for resource scarcity, uncertainties, and disasters in planning and designing the city have combined to create conditions that inhibit possibilities for just social reproduction and have caused socio-cide.

The flow of newcomers to Ramallah has been a traceable feature of the city's development since the turn of the 20th century (Aruri 2015a). Ramallah's expansion since the mid-1990s however, has been exponential and its population has more than tripled to over 40 000[6] persons (a total of about 125 000 persons are registered in the morphologically continuous Ramallah, alBireh, and Beitunya Metropolitan Area (Palestinian Central Bureau of Statistics 2019). This is due to several reasons, amongst which is being the seat of the Palestinian Authority (PA) and therewith its role as the 'green zone' (Massad 2006), and its relatively 'soft' social hierarchies, plurality of ideologies, and cultural liveliness. What has also turned acute in this century is, on the one hand, political and capital concentration in Ramallah (of, for example, PA institutions and diplomatic missions, companies and financial headquarters, and NGOs), and on the other, intensifying escalations in other urban and rural areas (including, for example, attacks by illegal Israeli settlers and incursions by the army, cuts in supplies of water and electricity, demolition of structures and farmlands), leading to real estate in Ramallah becoming the prime sector for speculative profits.[7]

Real estate speculation has been feeding into what Gregory calls an 'architecture of enmity'[8] (2004, p. 20) that materializes through intensifying contestation, urban spatial management patterns that pay little attention to public space provision, and forms of architecture that alienate people from each other and from the urban landscape; which are amongst the manifestations of spacio-cide. Therewith, contemporary developments have been contributing to inhibiting social interaction (such as co-learning and solidarity-building) and material sharing (be that land, or the concrete of the wall thus reducing costs and resource consumption). Besides the fact that most of the labourers in the construction field[9] are day-labourers with no insurance or benefits, Ramallah's urban growth patterns[10] have been increasing isolation and vulnerability whether to policing, or disasters, or simple everyday needs. As Anani (2011) states:

> The hegemony of exchange value makes the city mobile, movable and transformative like a container harbor, but at a slower rate. When a society loses its organic connection with its history, geography, climate, tradition, language and values, urban space becomes a purely functional setting, like the container, based primarily on utilitarian considerations that can be easily copied and transferred from one place to another. This containerization of urban space, and the fragmentation of the humane, essential for the production of urban space, has its root in the history of colonial policies fragmenting and separating the Palestinians from their geography, society, culture and environment. (n.p.)

During the 25 years since the Oslo Accords, when the colonial reality is still omnipresent but also when the PA has relative administrative freedom within Areas A (city and town cores – approximately 18% of the

West Bank and 4% of Mandate Palestine), there has been neither a comprehensive revision nor significant altering of planning and design regulations, for which the British Mandate (1919–1948) had set the codes. Acknowledging the mechanisms of colonization cannot be limited to overt symbols – for example, checkpoints, illegal Israeli colonies – but should equally investigate the roots of the continuation of these mechanisms and their contemporary applications in terms of engendering social, corporeal, and resource insecurities in the everyday urban spaces of Palestinians.

In this regard, Ramallah's municipal plans and strategies continue to reproduce the colonial intent of spatial militarization (surveillance, policing, and elimination of spaces where opposition can manifest), for which the current panoptical mobility system presents evidence, as seen in Figure 8.1. While routes for vehicles are rhizomatous (allowing the state to monitor and swiftly reach areas to repress opposition), pedestrian routes, which are fundamental for less-privileged residents, have severely shrunk in size and numbers. Likewise, Ramallah's Municipality has channelled significant resources since 2008 (mostly through ODA and Corporate Social Responsibility programmes) to the redesign of traffic crossings and *Mayadeen* (squares) (Aruri

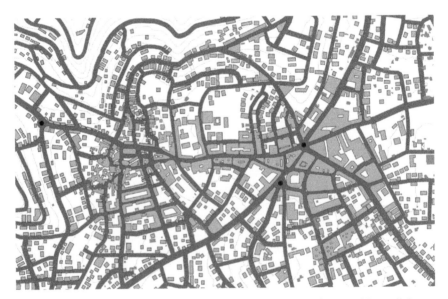

Figure 8.1 Illustration of the street network in a central part of Ramallah. In the past few years the network has extended well into the valleys and is covering much of the water-ways. Aruri 2015a

2015a). The name, Mayadeen, implies that these spaces will serve the public in exercising its right to congregate. Yet, these are predominantly traffic junctures and none of them provide space to pedestrians besides liminal, narrow sidewalks. Such arrangements eliminate the ability to accommodate social activities such as festivals, cultural events, and demonstrations, without the interruption of traffic (Aruri 2015a). The consequence of this planning regime is socio-spacio-cide through framing public interactions of encounter and exchange as exceptional rather than integral to the everyday in urban spaces; thereby, the fortressing of the hegemonic regime against citizens enacting commonness and building solidarities, which are pivotal for shifting to just social reproduction.

Equally indicative of the socio-spacio-cide logic in urban planning in Palestine, of socio-spatial dissection as opposed to creating connections and continuities, is that of land parcelization and construction typologies that foreclose the creation of small-scaled public spaces such as neighbourhood staircases and yards. Typically, in such spaces, citizens of varying backgrounds can interact and break down hegemonic social orders. This is an aspect of the anti-politics machine in city-making, whereby the quality of life in the built environment is not factored in, and therefore providing real estate investors with the freedom to develop a parcel of land irrespective of its socio-environmental contexts. Real estate investors and developers in Palestine are mostly local or expats. Those operating on a large scale are often closely affiliated to the PA (e.g. the Palestine Investment Fund), or descend from affluent families (e.g. al Masri family, owners of multiple real estate companies), or from the nouveaux-riche. Furthermore, observers and analysts estimate that the current number of empty apartments in Ramallah ranges from between 15 000 and 20 000, while construction is still ongoing and prices rocketing because real estate is ideal for money laundering as billing is not monitored (Hilal and El-Sakka 2015).

In hilly Ramallah, developers are constructing multiples of what the official Master Plan set as the parameters. For example, in residential areas classified as 'A', built structures are allowed a maximum elevation of four floors and a roof above street-level, yet developers are able to add any number of 'basement' floors. As demonstrated in Figure 8.2, the building to the right has a total of seven basement floors (the dotted line indicates ground-floor level), constituting a 140% increase in total built-up area as compared to a building on the opposite side of the street that has no floors below ground level. Taking topography into account in housing and neighbourhood design, which is fundamental to social reproduction as I explain below, would translate into higher execution costs. As designers, developers, and the municipality acquire more

Figure 8.2   A typical apartment building in Ramallah. The dashed black line marks the ground-floor level. The dark and light grey shading approximately marks the excavated volumes, that would have been spared had planning and construction laws considered social and environmental factors, and the lacking and needed pedestrian mobility infrastructures. Aruri 2015a

money from built space, attempts to change construction laws have not succeeded.[11] Architects and urban planners are the enablers and executers in the continuation of the status quo in which short-term, privatized gain is the baseline. Common well-being and creating safe and healthy habitats are released from being the responsibility of investors, and outsourced to a patchwork of services by underfinanced authorities and the selective uncoordinated agendas of NGOs.

In Ramallah, the unbuilt areas of a developed parcel are limited to the legally required 'retention' and constitute narrow strips of land on the four sides of buildings, what Trancik (1986, p. 1) coins as 'antispaces':

The usual process of urban development treats buildings as isolated objects sited in the landscape, not as part of the larger fabric of streets, squares, and viable open space. Decisions about growth patterns are made from two-dimensional land-use plans, without considering the three-dimensional relationships between buildings and spaces and without a real understanding of human behavior. In this all too common process, urban space is seldom even thought of as an exterior volume with properties of shape and scale and with connections to other spaces. Therefore what emerges in most environmental settings today is unshaped antispace.

These antispaces, which increase panoptical surveillance and fragmentation through walling, are either unused or fenced as private gardens for the use of tenants of adjacent apartments, or sometimes serve as car-parking areas, while children play mostly on the street.[12] As a result, neighbours in the same or adjacent buildings seldom meet, rarely have any exchanges that could help remind them of and foster commons and empathy, and remain strangers. This is particularly troubling as in Ramallah more than half of the population is neither born nor raised in the city, rather newcomers from other regions are continuously adding to the cultural mix of the city (Palestinian Central Bureau of Statistics [PCBS] 2010). Politicians expect residents of Ramallah (and other cities) to act as a harmonious unit at a time when isolation is the rule and spaces of co-habitation and assimilation are absent.

Whether rural or urban, pre-Mandate Palestinian localities were predominantly composed of continuous morphologies of back-to-back structures. Besides efficiency in the use of space and materials and climatological specifications, compactness also served the reduction of earthquake-induced torque on structural elements (Palestine lies on a tectonic fault-line[13]), and enabled thinking about the neighbourhood as a continuum. Today the application of knowledge about earthquakes is limited to materials and the design of individual structural elements, while the accumulative dimension – how adjacent buildings and their landscapes affect each other's stability – is absent.[14] From this perspective, the logic of socio-cide is reflected in indicators of urban development and welfare where the focus lies on ownership of or access to habitats, and material considerations of the level of security, durability, and structural soundness of these habitats is omitted. And yet, urban morphology is central to understanding spatial conditions for just social reproduction as it directly links to the ability to accumulate wealth through the durability of personal investments in securities such as housing. As demonstrated in Figures 8.3 and 8.4, this problem reiterates across Ramallah.

Besides considerations of social interaction, of earthquakes and climate changes, antispaces increase the risk of collapse of buildings and retention walls (that can be as high as 20 metres) because ponding storm water can lead to unsettling of foundations. Ramallah has increasingly suffered from flash floods in winters and water shortages in summers due to Israeli cuts in supply, which is reduced to only one or two days per week depending on the region.[15] Yet, dealing with storm water is limited

Figure 8.3   A view of Ramallah from alTireh Neighborhood (north-west) direction the center (looking eastwards) showcasing the mushrooming buildings and resulting antispaces. Aruri 2015a

Figure 8.4   A typical new neighbourhood in Ramallah. Notice the narrow canyons of antispaces between buildings, the size of excavations, and retention walls behind the buildings. Aruri 2015a

Figure 8.5 Stormwater puddling in the antispace of a residential building in Ramallah after a storm in winter 2013. Aruri 2015a

to the construction of (expensive, ODA dependent) subterranean channels to direct waters towards valleys where they are captured by Israeli infrastructures. The design of these networks is disconnected from that of the landscapes of individual lots, which due to extreme excavations are becoming hazardous entrapments of storm water, as can be seen in Figure 8.5.

While the responsibility for water supply is ostensibly that of the Palestinian Water Authority, it merely channels the rationed water sanctioned by the Israeli Authorities. It took until 2016 for the Ramallah Municipality to add the construction of rainwater collection wells to the criteria for a construction permit (it is noteworthy that until the late 20th century this was a normative practice). Yet the small size of new wells indicates that they are more symbolic than based on calculated needs, and monitoring of the execution of this infrastructure is eclectic.[16] At a time when women remain the main – if not sole – caretakers of households, such infrastructural shortcomings add to their burdens. For example, in summer all laundry has to be done on the day water is supplied rather than timed and distributed or re-scheduled in accordance with women's many other responsibilities.

Furthermore, under the pretext of hygiene and privacy, modernist and functionalist concepts of 'zoning' have led to increasing numbers of 'Residential Areas', where commercial activities do not receive licences to set up shop (hence further reducing spaces of socialization). Consequently, residents of such neighbourhoods have to commute to purchase groceries and reach services or work. This consumes fuel and time, which adds to the costs of livelihoods and extracted wealth. Cindi Katz calls this 'space-time expansion' (2001, p. 1224), where capitalism and spacio-economic restructuring lead to extending the physical arenas and terrain of social production and reproduction. As households typically own one car that is mainly for the use of a senior man, and in the absence of reliable and wide-reaching public transportation networks, this adds another layer of burden on women and increases the hurdles set in their way against upward socio-economic mobility. Equally, in cases of emergency, be it a snowstorm or armed aggression by the Israeli army or a military curfew impeding movement across city sectors, residents are hostage as to whether response teams and supply lines can reach the affected neighbourhoods.

Overall, the current construction laws in Ramallah not only reproduce spatial militarization but also enable the creation of antispaces, and both are components of spacio-cide. The current morphology of Ramallah is mutated, constituting an urban environment where social interactions are limited and the deleterious repercussions of disasters (natural and human-made) are multiplied rather than contained. As resilience is key to social reproduction, planners must design new kinds of environments that account for the uncertainties of climate emergency and earthquakes (embedding disaster-response into everyday spatializations and designs of infrastructures) rather than the current assumption of stability. Hence, when the goal of the billions of dollars of ODA is socio-economic development, stabilization, and increasing people's resilience, and as crises are the rule rather than the exception in Palestine, city planners, designers, and urbanists must shift the paradigms of their thinking about and practices in making cities, including in Ramallah.

Ramallah's Tomorrow: Between Individualisms and Commons

As Lefebvre (2009 [1974], p. 56) has noted, every change in discourse, whether societal or spatial, requires investment, 'there is no getting around the fact that the bourgeoisie still has the initiative in its struggle for (and in) space.' Notwithstanding this reality, the bourgeoisie (the local elite in postcolonial contexts) tend to avoid risks, preferring secure investments that often align with colonial interests (for the case of Ramallah see, for example, Abourahme 2009; Anani 2011). For Fanon, in geographies where 'enormous wealth rubs shoulders with abject poverty' (2008, p. 117), one of the main problematics is the inability of elites to conceive new worlds, because it 'is lacking ideas, because it is inward looking, cut off from the people, sapped by its congenital incapacity to evaluate issues on the basis of the nation as a whole' (Fanon 2008, p. 102). Since the role of elites is circumscribed to that of subsidiary managers rather than entrepreneurs of new solutions, addressing the negative impacts of antispaces on social reproduction and forging new, decolonial discourses of planning urban spaces in Palestine depends on mobilizing alternative imaginaries and resources to form a critical mass.

Just as cities are never end-products but rather spaces of dynamic processes and layering over time, I argue that Ramallah's antispaces – the undeveloped leftover spaces around buildings that account for close to a third of the city's area (Aruri 2015a) – are the prime spaces for interrupting the current system of domination in which 'personal success is based on collective destruction' (alKhalili 2012, interview 10 October 2012). While investors have long moved to the next empty lot to maximally develop and sell for a fast return of profits, tenants of finished buildings – who can be seen as small-scale investors (86% of housing units are owned by a member of the household (Palestinian Central Bureau of Statistics 2018)) – are left to deal with the imperfections and daily hurdles.[17] Just as shock and poverty serve as entry points for disaster capitalism (Klein 2007), manifold daily difficulties and vulnerabilities due to mal-design and underuse of citywide antispaces are ecosystems incubating the social reproduction of non-opportunity.

And yet, antispaces could become an entry point to negotiate and create missing and much needed public areas and to reduce corporeal and economic precarity. Rethinking existing small and medium-sized enclosures as legitimate zones for public use can serve a number of purposes: i) re-adapting antispaces can improve living conditions and reduce disaster risk; ii) showcasing possible, low-cost models for higher efficiency in use of resources and antispaces can influence future forms of architecture; and iii) creating spaces for public, semi-public, and semi-private interactions to counter increasing otherness and contestation can increase the visibility of women's contributions in the public realm. People-based initiatives in antispaces can create empathy, solidarity, and recreate the pre-Oslo practice-based understanding of the responsibility of individual actors in creating (relative) collective well-being under colonial realities. While conceived by the colonial logic of 'divide and conquer', antispaces could become the very spaces of decolonization through recreating security valves that can absorb increasing economic pressures and counter segregation, and over time, shift urban design and planning paradigms to incorporate considerations for just social reproduction as a tool of resistance of colonization.

Changing construction laws in Ramallah so that antispaces can be regarded as connecting webs and enclosures of semi-public spaces would entail lengthy legislative reform that can take years and considerable costs. Such processes include detailing safety parameters, adding sub-clauses for ownership and privacy rights, and amending modes of decision-making, amongst other issues. Moreover, this would involve working against the grain of dominant neoliberal market logics which favour the current individuation of ownership and ability to develop lots independently. Furthermore, for average Ramallites, the current developmental patterns in terms of the spatial allocation and use of land have become the norm, whereby a public space is perceived as an external enclosure and almost never associated with the proximate environment. The lack of public space is perceived as a problem, the answer to which lies in local authorities acquiring and allocating spaces for parks, playgrounds, and similar leisure spaces. Everyday spaces of encounter, such as a neighbourhood plaza, is not part of citizens' vocabulary, and is not seen as an element that affects social reproduction. For most citizens, the idea of having 'strangers' strolling along the side or back façades of their home is considered an invasion of privacy, even though they do not use these spaces and tend to keep their shutters and curtains closed to avoid residents in adjacent buildings, a mere six or eight metres away, from being able to see in. Hence, allowing the abundant antispaces to be available for shared use, to vitalize them as spaces of possibility, opportunity, and reduction of inequalities, will not happen without a shift in mindset. Amongst other aspects, this means the de-individualization of security,[18] and reestablishing the pre-Oslo social contracts whereby citizenship (as belonging) concerns both rights and responsibilities.

The way forward is not through the bureaus of planners, but rather, through makeshift civic-based experiments[19] that, as Brugmann (2009) points out, can create a positive coalescence between public good and private benefit. People have to be able to see and touch alternative realities that they do not imagine possible, without the fear that the new propositions are permanent, imposed realities. People have to be able to see themselves as the architects in the reproduction of their lived spaces, and that their daily social engagements are integral to overall welfare – the safety nets that buffer capitalist hegemony.

Coined by Katherine Shonfield, *premature gratification* describes such engagements that are propositional and locational, seeking long-term consequences, while constituting 'brief disobedience' undertakings as a 'means to advance proposals in advance of advancing proposals' (Thomas 2014, p. 150). In the case of Ramallah, reconfiguring selected antispaces using makeshift arrangements can exemplify alternatives to current spatial planning regulations. For Ferguson (2014), urban interventions of this nature provide a platform for urgent re-negotiations on resource-allocation, modes of economic production, and social and political principles in both real and symbolic spaces. While 'real spaces' are acquired in the interim materialization of alternative spaces, the virtue of such make-shift spatial experiments becoming 'symbolic spaces' depends on their success in forging new solidarities and modes of city-making.

Planning in Ramallah features a mixture of public and private instruments that produce both inclusive spaces, such as urban playgrounds, and exclusive spaces, such as the gated neighbourhood of the Diplomatic Quarter. However, 'commons' as productive spaces that are jointly owned and governed are lacking in today's urbanized Ramallah. As Harvey (2013, p. 71) notes, commons are the result of social practices and political action by citizens who have to decide 'whose side are you on, whose common interests do you seek to protect, and by what means?' Harvey also highlights that the rise of commoning as an alternative is a Cartesian product of the scarcity of public goods orchestrated by neoliberalism. Hence, when propagating insurgent urban commons as antidotes to disenfranchisement and for changing systems of social reproduction, we should keep in mind that '[u]nfortunately the idea of the commons (like the right to the city) is just as easily appropriated by existing political power as is the value to be extracted from an actual urban common by real estate interests' (Harvey 2013, p. 87).

Refiguring and Reconfiguring for Resilience: Takhayyali [Imagine] Ramallah

One way in which the decolonization of Ramallah can move forward is by reintroducing the Masha' (shared land ownership, i.e. a common) with a contemporary translation of its function to the urban, particularly in antispaces. In agriculture, this tool enabled farmers to prevent soil degradation by rotating various crop-types along lots, and through its governance structures it socialized the costs and risks of this tough practice, thereby protecting individual farmers against collapse. This tool is still part of existing legal systems for land management and can be employed in the urban context in combination with concepts of neighbourhood (self)management.

There are various experiments worldwide in formal and informal neighbourhood management systems such as Quartiersmanagement in Berlin (Colomb 2012), the grassroots-based favela-barrio projects in cities in South America, and those by Slum Dwellers International in Africa and Asia (Lines and Makau 2018), amongst others. While each experiment has had different contextual components, dynamics, and rates of (non)success, what underlines all of them is the focus on creating social cohesion as a pillar for dealing with vulnerabilities and the creation of security nets. They establish community centres (mostly voluntary-based) that: provide education and training in self-help and labour skills; coordinate activities that include infrastructural repair; provide waste collection and re-/up-cycling; and create urban gardens that provide basic supplies of food and education about produce cycles, amongst other activities. In place of waiting for government or NGO solutions, they communicate results to local authorities as evidence to push for the reform of policies and legislation, towards creating better chances of access to opportunity for the marginalized. In a sense, what such initiatives have been forging is what Hall calls 'policymaking as social learning', where '[p]olitics finds it sources not only in power but also in uncertainty – men [sic] collectively wondering what to do' (1993, p. 275). So what can be done in Ramallah and how would the Masha' be employed to create urban spaces that cater for more just social reproduction?

Master plans in Ramallah are mostly prepared in two-dimensions (horizontal), with limited consideration to the third (vertical) dimension. When rethinking antispaces, a makeshift experiment can aim at re-imagining city spaces in terms of varying layers of ownership and management that shift along the X, Y, and Z planes, as well as in the fourth dimension: time. While built-up spaces with residential, commercial, or other functions can continue to be privately owned and governed, theoretically, it is possible that antispaces of multiple adjacent lots can be re-defined as Masha', whereby residents and owners have shares in the overall and not only the individual lot. In Figures 8.6a and 8.6b, I use the shading to illustrate the antispaces that could be re-thought as collective ownership, where complimentary functions could be programmed.

Under such a scenario, residents would have a stake and a vote in the uses, designs, management, maintenance, and upgrading of the antispaces around their residential units and further within the block or neighbourhood. Such a reconfiguration of thinking, less about the individual parcel and more about the neighbourhood as continuum, is important to create contiguous ecosystems and greater efficiency in the use of resources.

It could break the mental and physical isolation of citizens by creating mediums for quotidian interaction, and nurture solidarity through shared caring for and benefitting from the common. In this sense, such steps would amount to what Harvey describes as identifying political and social alternatives to operations 'of the capitalist law of value' within the market (2013, p. 123), which is necessary for creating new norms for systems of social reproduction.

Thus, in place of each lot separately funding uses that make up for the absence of state-provisioned public spaces and social infrastructures (for example, a public staircase, a water well, a small play area for children, or a herb garden), functions can be bundled amongst several neighbouring lots. This would, first, reduce and socialize the costs of implementation, making investments feasible for average citizens (not only affluent and rich developers). Second, the resulting spaces and designs would be more satisfactory to local needs and desires (e.g. levels of privacy controls and accessibility of non-residents of the block to developed antispaces) as they are determined by and with the residents for the residents, rather than by subjective professionals, such as, for example, architects who do not necessarily live in the area. And third, such spaces could be a way to address the structural barriers such as, for example, the lack of affordable, accessible, and proximate public spaces and high land prices, which make implementation of small neighbourhood social projects not feasible, and more women into public life from which they are currently increasingly excluded.

This approach works with the dominant neoliberal city management strategies that have been pushing for the decentralization of services and their privatization under the pretext of efficiency through relieving pressure on limited municipal capabilities and reducing bureaucracy. Hitherto, these strategies have only been applied to move responsibility from civil servants to weakly coordinated, profit-oriented companies under the framework of Public Private Partnerships (PPPs), through which implementing parties acquire subsidized access to valuable public resources under vague standards and accountability parameters. As Healey et al. (1997) pointed out decades ago and as substantiated by several others since, the result has been city development patterns in which the responsibility for creating compatible hardware (e.g. buildings, streets, public spaces) and software (e.g. social infrastructures, service supplies) systems that produce functioning, holistic urban realms is lost, and dependency and vulnerability are entrenched.

What is needed in Ramallah is a generation of urbanists that can imagine alternative realities and systems that break from the dominant

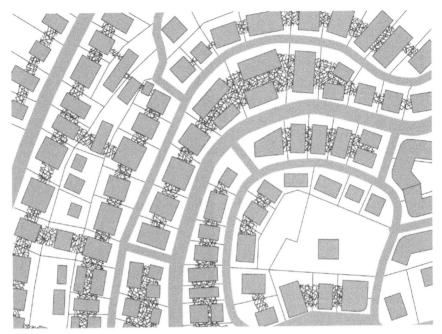

Figure 8.6a The first of two figures showing a schematic illustration of a central neighborhood in Ramallah. In this figure the spaces demarcated with hatching (between buildings) are areas that can be set into use for semi-/public purposes using existing laws and regulations, if neighbors agree to cooperate with each other. Aruri 2015a

paradigms that are dissecting city spaces, and where wealth extraction by capitalists is eased and risk and burdens are loaded onto everyday citizens, particularly women. We need urbanists who can communicate their ideas to inspire wider re-figurations of city spaces, and 'unionize' residents into neighbourhood management schemes that can voice their collective claims for, and materialize their desires for, just city spaces. We should acknowledge that we do not have the needed solutions, and the only way to find them is through experimentation and incremental learning. As Hall and Jones (2011) remind us, in all sectors including the urban, policy-making is a process of collective puzzling that operates through continuously growing sets of flexible parameters and conditionalities. Hence urbanists should delink their work from its focus on the setting of master plans (which consume considerable time and money, are inflexible to the rapidly changing realities, are based on assumptions of stability, and are not known to account for issues such as gender and intersectionality), and envisioning

Figure 8.6b The Second of two figures showing schematic illustrationof a central neighborhood in Ramallah. This second figure is an ambitious forecasting into the future, what could possibly materialize as reclaimed antispaces if the urban Masha' concept gains shape and popularity. Aruri 2015a

particular aesthetics that are bound to change with technological and material advancements. Rather, as Johar (2014) implores, architects, urbanists, and city authorities should start investing more time in thinking about urban planning and policy-making as curating non-centralized processes of city re-/making, that lead to cross-sectoral capacities in dealing with disenfranchisement, dispossession, uncertainties, and disasters.

Particularly, in terms of dealing with risk and disasters, experience has shown that the presence of proactive and networkable/-ed local communities is crucial for responding to daily needs as well as sudden shocks, particularly in deterring 'disaster capitalism' from capturing more ground (Klein 2017). Hence urbanists should find ways to break from the anti-politics machine, from business as usual, and acknowledge that they are needed for conceiving new social reproduction environments. They are the ones most capable of creating windows in city-making systems that give space for communities, particularly women, to test initiatives and apply ideas that nurture and enable resilience against crises and disasters, and to help reconfigure everyday behaviours such that they bring about sustainable, equitable modes of living.

In this chapter, I have named an entry point for current and future generations of Ramallite designers, planners, and spatial entrepreneurs to test their imaginations, and shown how they can play a role in refiguration of our city towards decolonization. Antispaces can be reconfigured in many different ways to host singular or multi-functions that improve life quality, provide people-based safety nets, and inspire new modes of policy-making that centre around what brings people together rather than what differentiates them. Some antispaces can be reconfigured to create spaces for social uses such as courtyards, plazas, and playgrounds for children. Some can be reconfigured to serve water capturing, filtration, and re-use. Some can be reconfigured to install funicular railways such as in Valparaíso, Chile, to offer an alternative to vehicle-based mobility in the hilly landscape. Some can be reconfigured to host horizontal and vertical urban farms that provide local employment, raise the consciousness of people to how long and resource-extensive growing produce is, and where, for example, the rent of the antispace goes to residents to increase and diversify their income sources. These and other ideas could help in giving visibility to women's bodies and labour, alleviate burdens, and reduce hurdles against their upward socio-economic mobility. As such, rethinking antispaces is a step towards reconfiguring the dimensions through which social reproduction operates within and alongside commodity and spatial productions, and where urban space is examined as a continuum rather than parcelled. Urban space should be acknowledged as a medium of materialization, refiguration, and re-establishing of relationalities and connections. As Katz argues, since 'topographical knowledge is so integrally important to capitalists and other agents of domination and to the maintenance of uneven development, its appropriation should be important to countering them' (2001, p. 1215).

For these ideas to transcend being sporadic virtual refigurations and to become widespread material reconfigurations, it must be shown that they can lead to a common good rather than new extractive processes. An entry point to open conversations in Ramallah can be the long overdue talk on disaster risk reduction and spatial safety measures, which touch on the role of the individual in building resilience and response capabilities, preemptively and in the aftermath of disasters. If this succeeds, then we might be one step closer in forging alternatives and guarantee better access opportunities for average Ramallites and in turn, decolonization. And if that happens in Ramallah, maybe we will be able to showcase a bottom-up motion towards real sustainability, that transcends global political agendas to local realities. As Lefebvre (2009 [1974], p. 54) tells us:

> A revolution that does not produce a new space has not realized its full potential; indeed it has failed in that it has not changed life itself, but has merely changed ideological superstructures, institutions or political appa-ratuses. A social transformation, to be truly revolutionary in character, must manifest a creative capacity in its effects on daily life, on language and on space.

Acknowledgements

This article is based on research from my Ph.D. thesis and was produced within the framework of the research project 'Urbanization, Gender and the Global South: a transformative knowledge network', which is supported by the Social Sciences and Humanities Research Council of Canada (SSHRC) grant number: 895-2017-1011.

Notes

1 The Oslo Accords of 1993 divided the West Bank and Gaza Strip into: Areas A (under administrative and policing control by the Palestinian Author-ity); Areas B (under the administrative control of the Palestinian Authority but policed by the Israeli Army); and Areas C (under full Israeli control). Today, Area C covers 60% of the total area of the West Bank.

2 This research was conducted between 2010 and 2014 and based on: 20 semi-structured qualitative interviews with experts and profession-als from various backgrounds (public officials, private investors, urban-ists and planners, politicians, economists, journalists and researchers); 8 focus groups; and analysis of legal documents, public announcements, market operations, quotidian comments on social media, and photo documentation.

3 Examples of social contracts in pre-Oslo times include: the Popular Vol-untary Work committees that had thousands of members (mostly youth) carrying out public works and services that are normally the responsibility of ruling authorities and in this case the Israeli government; the Society of Inash alUsra; the Palestinian Agriculture Relief Committees; the Medical Relief Committees; and the Palestinian Students Union. Amongst many

other examples, there is the cow-milk farm in Beit Jala whose story is documented in the animated documentary 'The Wanted 18'.

4   For example, as one of its strategic goals, the Ramallah Municipality is seeking the establishment of an Expo Center and has allocated prime territory and significant resources for it. Yet, the question of mobility, whether of people from outside the city or of goods (e.g. to be exhibited), remains in the hands of the Major-General serving as Commander in Chief of the Israeli COGAT (Coordination of Government Activities in the Territories). The everyday reality of the past two decades has been at best a partial closure of borders and at worst (and often) a total curfew.

5   See note 3.

6   Observers estimate that the actual number of persons living in the city far exceeds 40 000, where many of those who moved to Ramallah maintain their registration elsewhere, because the taxes in the city are the highest amongst all Palestinian localities.

7   Investments in real estate are very high because productive economic sectors are hampered by the Israeli strategies to maintain Palestinian dependency.

8   The 'architecture of enmity' refers to physical representations that 'give shape and substance' to fears and 'inhabit dispositions and practices, investing them with meaning and legitimation' towards cementing the 'difference of others' through spatial dynamics and aesthetics (Gregory 2004, p. 20).

9   According to the Palestinian Central Bureau of Statistics Census of 2017, persons (aged 15 years and above) employed in the construction sector account to almost 20% of the total number of employed persons (PCBS 2019).

10   For information on the nature of the real estate sector and the expansion of Ramallah, see Aruri 2015a.

11   Construction laws and regulations are the responsibility of the Ministry of Local Government and the Higher Planning Council. This has made it easier for the Municipality of Ramallah (as well as others) and the Chamber of Engineers to evade tackling the persisting problems of current construction codes as they officially lie beyond their legal jurisdiction, in spite of the fact that they can present claims if they wished to.

12   Although the Municipality of Ramallah has been landscaping some small neighbourhood gardens, these remain insufficient in number.

13   In terms of geology, the fault-line of the African and Arabian tectonic plates runs along the north-south axis of Palestine, marking the four regional earthquake provinces. Formerly, the compactness of built environments in Palestine served to reduce earthquake-induced torque on structural elements by statically reframing it into a linear momentum that is in turn curtailed through Newton's third law of motion. For detailed analysis, see Aruri 2015b, pp. 60–72.

14   The higher the building, the larger the torque and structural failure possibilities. Besides torque curtailment, in the event of structural failure, sections of detached buildings would collapse irregularly in any direction. Such disintegrating sections are thence momentarily transformed into projectiles with high momentum. The same applies for retention walls.

15   Palestinians are not allowed to pump water out of the aquifers or construct large-scale water capturing infrastructures. Water has to be purchased from Israeli companies at four times the price offered to Israeli consumers. In summer it is customary that Israeli companies apply cuts, whereby water volumes are rationed and it is supplied irregularly on one or two days per week only.

16   Place-based systems for separation, filtration, and re-use of grey and black water are not yet part of municipal strategies to deal with the growing water crisis.

17   Besides those described earlier in this chapter, daily hurdles also include the climatic problems of over-heating and over-cooling, accelerating winds through the canyons between facades of massive building, the lack of safe spaces for children to play, and the taunting reality that in the absence of suitable pedestrian walkways, although the supermarket lies only a 100 metres away, nonetheless one needs to travel about a kilometre up- or down-hill to reach it. And when cars are blocked in due to snowstorms, municipal response teams add delivery of supplies to their urgent duties.

18   For example, depending on local shopkeepers and shoppers on the street to deter thieves from buildings in place of expensive home security devices; or sharing the burdens and benefits of garden-grown kitchen herbs in place of their purchase from vendors, etc.

19   Make-shift experiments resemble in some aspects 'tactical urbanism', 'DIY urbanism', and 'Guerilla urbanism'. It is temporary in nature and does not seek to create permanent realities, but amongst its aims is to inspire potential longer-term spatial reconfiguration.

References

Abdo, N. (2014). *Captive Revolution. Palestinian Women's Anti-Colonial Struggle within the Israeli Prison System*. London: Pluto Press.

Abourahme, N. (2009). The bantustan sublime: Reframing the colonial in Ramallah. *City* 13 (4): 499–509.

Abourahme, N. (2011). Spatial collisions and discordant temporalities: Everyday life between camp and checkpoint. *International Journal of Urban and Regional Research* 35 (2): 453–461.

Al-Haq (2015). Violence against Palestinian women must stop. *Al-Haq* (2 December). www.alhaq.org/advocacy/topics/palestinian-violations/1001-violence-against-palestinian-women-must-stop (accessed 15 August 2018).

alKhalili, Y. (2012). Urban Development of Ramallah post the signature of Oslo Accords. *Interviewed by Natasha Aruri [in person] Ramallah*, 10 October 2012.

Anani, Y. (2011). Academia, the mirror. In: *Proceeding of the Symposium Designing Civil Encounter*, Ramallah (21–24 July 2011). www.artterritories. net/designingcivicencounter/?page_id=13 (accessed 13 May 2012)

Appadurai, A. (2003). Sovereignty without territoriality: Notes for a postnational geography. In: *The Anthropology of Space and Place: Locating Culture* (eds. S. Low and D. Lawrence), 337–349. Oxford: Blackwell Publishing.

Appadurai, A. (2008). *Modernity at Large: Cultural Dimensions of Globalization*. Minneapolis, MN: University of Minnesota Press.

Aruri, N. (2015a). *100 × Ramallah: Imaginations, otherness, and (de)colonization in antispaces of sumud, 1914–2014*. Doctoral Dissertation. University of Duisburg Essen.

Aruri, N. (2015b). Rediscovering little sins: Palestinianhood, disobedience, and Ramallah. *Rosa Luxemburg Stiftung PAL Papers* (November). www. rosaluxemburg.ps/wp-content/uploads/2015/12/Rosa-Luxemburg-Articles-English-Natasha-Aruri-Paper.pdf (accessed 10 November 2018).

Bakker, I. and Gill, S. (2003). Ontology, method, and hypotheses. In: *Power, Production and Social Reproduction. Human In/security in the Global Political Economy* (eds. I. Bakker and S. Gill), 17–42. New York: Palgrave Macmillan.

Brugmann, J. (2009). *Welcome to the Urban Revolution, How Cities Are Changing the World*. New York: Bloomsbury Press.

Colomb, C. (2012). *Staging the New Berlin, Place Marketing and the Politics of Urban Reinvention Post-1989*. Oxfordshire: Routledge.

Fanon, F. (2008). *Black Skin, White Masks*. New York: Grove Press.

Ferguson, F. ed. (2014). *Make_Shift City: Renegotiating the Urban Commons*. Berlin: Jovis.

Ferguson, J. (2009). *The Anti-Politics Machine: Development, Depoliticization, and Bureaucratic Power in Lesotho*. Minneapolis, MN: University of Minnesota Press, Minneapolis.

Ferguson, S. and McNally, D. (2014). Precarious migrants: Gender, race and the social reproduction of a global working class. *Socialist Register* 51: 1–23.

Graham, S. ed. (2006). *Cities, War, and Terrorism, Towards an Urban Geopolitics*. Third Edition. Oxford: Blackwell Publishing.

Gregory, D. (2004). *The Colonial Present*. Oxford: Blackwell Publishing.

Hall, P. (1993). Policy paradigms, social learning, and the state: The case of economic policymaking in Britain. *Comparative Politics* 25 (3): 275–296.

Hall, P. and Jones, M. (2011). *Urban and Regional Planning*. Fifth Edition. New York: Routledge.

Hanafi, S. (2009). Spacio-cide as a colonial politics in Palestinian Territory. *Grotius International, Géopolitiques de l'Humanitaire* (30 September).

www.grotius.fr/spacio-cide-as-a-colonial-politics-in-palestinian-territory (accessed 2 February 2014).

Harvey, D. (2013). *Rebel Cities. From the Right to the City to the Urban Revolution*. London: Verso.

Healey, P., Cameron, S., Davoudi, S. et al. (1997). *Managing Cities*. Chichester: John Wiley & Sons.

Hilal, J. and El-Sakka, A. (2015). *A Reading on the Socio Urban Changes in Ramalah and Kufur Aqab*. The Center for Development Studies. Birzeit University, Birzeit.

Johar, I. (2014). Architecture of the civic economy. In: *Make_Shift City: Renegotiating the Urban Commons* (ed. F. Ferguson), 204–207. Berlin: Jovis.

Katz, C. (2001). On the grounds of globalization: A topography for feminist political engagement. *Signs* 26 (4): 1213–1234.

Klein, N. (2007). *The Shock Doctrine: The Rise of Disaster Capitalism*. London: Penguin.

Klein, N. (2017). Address to the Labour Party Conference 2017. *YouTube* (26 September). www.youtube.com/watch?v=jj1nuw38DqY (accessed 1 October 2017).

Lagerquist, P. (2003). Privatizing the occupation: The political economy of an Oslo development project. *Journal of Palestine Studies* 32 (2): 5–20.

Lefebvre, H. (2009 [1974]). *The Production of Space* (trans. D.N. Smith). Second Edition. Oxford: Blackwell Publishing.

Lines, K. and Makau, J. (2018). Taking the long view: 20 years of Muungano wa Wanavijiji, the Kenyan federation of slum dwellers. *Environment and Urbanization* 30 (2): 407–424.

Massad, J. (2006) Pinochet in Palestine. *Al-Ahram Weekly* (9–15 November). www.weekly.ahram.org.eg/2006/819/op2.htm (accessed 20 April 2014).

Palestinian Central Bureau of Statistics (2010). *Localities in Ramallah & Al Bireh Governorate by type of locality and population estimates, 2007–2016*. www.pcbs.gov.ps/Portals/_Rainbow/Documents/ramallah.htm (accessed 20 April 2017).

Palestinian Central Bureau of Statistics (2018). *Press release by the Palestinian Central Bureau of Statistics (PCBS) on the occasion of Arab Housing* -Palestinian Central Bureau of Statistics (2019) *Population final results – Detailed report West Bank – Population, housing and establishments census 2017*. www.pcbs.gov.ps/PCBS_2012/Publications.aspx?CatId=26&scatId=0 (accessed 1 August 2019).

Samara, A. (2000). Globalization, the Palestinian economy, and the 'Peace Process'. *Journal of Palestine Studies* 29 (2): 20–34.

Taraki, L. (2008). Enclave micropolis: The paradoxical case of Ramallah/Al-Bireh. *Journal of Palestine Studies* 37 (4): 6–20.

Tartir, A. (2017a). The EUPOl COPPS Operations in Palestine: Aiding and abetting authoritarianism? *YouTube* (15 October). www.youtube.com/watch?v=epYvqXXQ_XU (accessed 18 November 2018).

Ta rtir, A. (2017b). The Palestinian Authority Security Forces: Whose security? *Al-Shabaka* (May 16). www.al-shabaka.org/briefs/palestinian-authority-security-forces-whose-security (accessed 18 November 2018).

Thomas, H. (2014). Brief disobedience and premature gratification. In: *Make_Shift City: Renegotiating the Urban Commons* (ed. F. Ferguson), 150–153. Berlin: Jovis.

Trancik, R. (1986). *Finding Lost Space: Theories of Urban Design*. New York: Van Nostrand Reinhold.

UNISDR (2017). What day is October 13th? *YouTube* (12 October). www.youtube.com/watch?v=5MtAwCe8CQk& (accessed 13 October 2017).

9

Forced Displacement, Migration, and (Trans)national Care Networks

Practices of Urban Space Production in Colombia and Spain

Camila Esguerra Muelle (National University of Colombia)
Diana Ojeda (Universidad de los Andes)
Friederike Fleischer (Universidad de los Andes)

Introduction

Over half a century of war has situated Colombia as the country with the highest level of internally displaced people in the world. Official records account for approximately 8 million people who have been forcibly displaced from rural areas to urban and peri-urban centres (International Organization for Migration (IOM) 2018).[1] Social and armed conflict, and a post-conflict scenario largely defined by dispossession dynamics associated with the expansion of agribusiness, mining, and other forms of extractivism, have historically produced rural areas and urban peripheries in Colombia through geographies of expulsion. Despite the recent peace agreement in 2017 celebrated between the government and

A Feminist Urban Theory for our Time: Rethinking Social Reproduction and the Urban, First Edition. Edited by Linda Peake, Elsa Koleth, Gökbörü Sarp Tanyildiz, Rajyashree N. Reddy & darren patrick/dp.
© 2021 John Wiley & Sons Ltd. Published 2021 by John Wiley & Sons Ltd.

FARC (Revolutionary Armed Forces of Colombia) guerrillas, internal forced displacement is increasing in the country (UNHCR 2017). Moreover, during the last three years, due to the Venezuelan crisis, Colombia has become the host country for over a million people coming from its neighbouring country, including Colombian returnees (Migración Colombia 2019).

Drawing on feminist and *descolonial* perspectives, our research examines the case of Colombia in order to expand our understanding of the connections between forced displacement, international migration, and the configuration of (trans)national care networks in relation to the urban. (Trans)national care networks impel precariously placed people – and particularly women and other feminized subjects – to survive through their articulation with informal economies. They generate wealth for corporations, as well as all forms of mafias, in addition to producing returns for national economies through the flow of remittances and currency. By centering our attention on paid care work, we seek to highlight migrants' important role in social reproduction and point to how the restricted spaces they occupy in the city reveal uneven geographies of urban production fuelled by exploitation, marginalization, and the mobilization of fear.

Focusing on the life trajectories and experiences of Colombian displaced people and migrants who carry out paid care work in Colombian and Spanish cities, we analyse the ways in which they participate in contested processes of the production of urban space, in both the public and private realms. As we show throughout this chapter, their testimonies can be read as examples of the way in which neoliberalism, revanchism, and hypervigilance shape urban space. Careworkers' precarious working conditions, criminalization, and reduced space for manoeuver are played out in the always gendered, situated, embodied, and contested practices of the production of urban space. Both insiders and outsiders of the cities they end up inhabiting, displaced people's and migrants' role in the production of urban space has received little attention in the Colombian context and needs further examination. We believe this is even more so for the case of women and people who do not conform to the mandates of cisheteronormativity, given the prevalence of violence in their daily lives. Official numbers, a clear underestimate, do not give account of the connections between migration and gender-based violence. In Colombia, most people identified as LGBT who are registered as victims of the armed conflict have been forcedly displaced (3105 out of 5097 according to [UARIV 2019]), and the multiple cases of violence to which they are subjected make it difficult to consider their migration as only voluntary.

From a collaborative multi-sited ethnography carried out in four Colombian cities – Cali, Cartagena, Bogotá, and Medellín – and two Spanish cities – Madrid and Barcelona – with 135 cisgender women and transgender people, we examine how (trans)national networks of care contribute to the production of urban space. Our research examines how the intertwined dynamics of war and globalized capital have forged a problematic geography of care work through which colonial power is constantly re-enacted. It pays particular attention to the ways in which women and other subjects without the privileges of whiteness, masculinity, ableness, or heterosexuality become *migration fodder* (Esguerra Muelle 2019), i.e. sacrificial bodies and existences, which can be commercialized and trapped into the frontier industrial complex as *maquila* workers, domestic servants, and cheap care work providers, amongst other marginal forms of employment. We explore here how these subjects become trapped in a cycle of migration-return and are effectively disposed to sustain (but not without contestation) uneven processes of urban production. We ask, how do these (trans)national care networks articulate the rural with the urban, at the expense of rural spaces, impoverished populations, and the global South, while producing urban spaces nationally and overseas?

The chapter proceeds as follows. We first outline the epistemological and methodological foundations of our research, pointing to how feminist descolonial knowledge can contribute to a better understanding of the relations between war, social reproduction, (trans)national care networks, and the urban. Reflecting on the experience of migrant careworkers in Spain, the following section remarks on the centrality of war and multiple, relational forms of violence in the configuration of (trans)national networks of care. From a feminist perspective, we argue that war cannot be reduced to an external cause of migration. Instead, we show its centrality to the configuration of careworkers and their place within cities like Madrid and Barcelona. We then focus on the work of *madres comunitarias* (communitarian mothers) in Colombia, who conduct a form of underpaid care work sustained mostly by women of rural origin who have been forcedly displaced to various cities in Colombia. In concluding, we point to the ways in which a feminist descolonial approach can contribute to the comprehension of processes and practices of the production of urban space.

(Trans)national Care Networks, Social Reproduction, and Urban Space

We refer to social reproduction as a series of apparatuses, relationships, dynamics, arrangements, and resources that sustain life, that is, that guarantee the material, symbolic, and emotional perpetuation of society. It includes practices such as procreation, child rearing, socialization, caring for life, and dealing with death. Social reproduction implicates a complex interweaving of power relations that results in both the conservation of and struggles against the social, economic, cultural, and political order produced in and through domestic spaces and beyond, which cannot be reduced to the reproduction of the workforce and capitalist relations (Mitchell, Marston, and Katz 2004; Meehan and Strauss 2015). Understood as such, care work is central to social reproduction. As a field of work that is devalued, thus rarely and poorly paid care work has been usually assigned to women, particularly to rural and migrant women, and those marked racially or by their non-normative gender identity and sexual orientation (Esguerra Muelle, Sepúlveda Sanabria and Fleischer 2018). In this sense, care regimes refer to historically, culturally, and geopolitically situated systems of care provision that are also political, discursive, and ideological.

Inspired by feminist conceptualizations of care work (Arango 1993, 2018; Ungerson 1999, 2004; Leonard and Fraser 2016; Klammer and Letablier 2007), we analyse an increasingly (trans)national care regime in which war dynamics in Colombia are entangled with the reproduction of cities such as Bogota and Madrid through forced displacement and exile. We do this from an intersectional perspective that understands care in terms of a network woven through relations of interdependence (Tronto 2013, 2016) and that reveals the centrality of care in maintaining interlocked systems of oppression marked by gender, age, race, class, sexuality, ethnicity, and ability (Dorlin and Bidet-Mordrel 2009; Carbado et al. 2013; Viveros Vigoya 2016). Esguerra Muelle's (2014) previous research has pointed to the important role of migration in the configuration of (trans)national care regimes. Using the notion of 'global care chains', Hochschild (2000) has shown how the precarious work of women from countries in the global South covers the care deficit of global North countries: the 'care drain' occurs through migration to enable the reproduction of value chains in a post-Fordist system of production (Bettio, Simonazzi, and Villa 2006). We seek to complicate the linearity behind this notion of 'global chains of care' by further delving into the (trans)national entanglements that constitute complex care networks, and by pointing to the effects of these entanglements on the everyday life of careworkers.

We propose an understanding of (trans)national care networks based on the Spanish translation as *tramas (trans)nacionales de cuidado*. As Esguerra Muelle (2019) suggests, the different meanings of the word *trama* can help us address the complexity of care work and its (trans)national dimension. The trama of feminized migration networks can be understood as: i) a set of strings, which crossed and knotted, form the warp and weft of a fabric; a network of affections, emotions, and sensations inscribed on the body of careworkers; ii) a narration which connects different elements of the stories and experiences of migrants; and iii) a metaphor of a confabulation or plot that causes harm, signalling the different forms of violence involved in the constant configuration of these (trans)national networks of care. In what follows, we show how (trans)national care networks – as fabric, narration, and plot – participate in the production of urban spaces in Colombia and Spain.

Based on our empirical work in Colombia, we show both the trans- and intra- national character of these networks and the ways in which forced displacement and exile has become an important source in the production of what Esguerra Muelle (2019) refers to as 'migration fodder', that is, the way in which these networks entrap women and other feminized subjects as careworkers in perverse cycles of migration within and beyond national borders. Focusing on migrant careworkers' experiences and positionalities within (trans)national and intra-national care networks, our research is based on both a multi-sited (Marcus 1995, 2018), decade-long collaborative ethnography carried out by Esguerra Muelle between 2008 and 2018 and Ojeda's (Ojeda and González 2018) and Fleischer's (Fleischer and Marín 2019) research on the experience of peasant women who have fled from rural areas in the context of war. Esguerra Muelle's research was conducted with domestic migrants living in Cartagena, Bogotá, Medellín, and Cali, in Colombia, and international migrants living in Barcelona and Madrid, in Spain. These subjects carried out paid domestic labour and childcare, as well as other forms of care work such as being nurses, hotel cleaners, and sex workers. Most of the participants were women of rural origin, although the research included people with a variety of ethnic-racial, sexual, and gender identities, and with different ages and abilities, with particular attention paid to the experiences of lesbian, queer, and trans women and men.[2]

Following critical feminist work on migration and care (Bakker and Silvey 2008; Parreñas 2012), we delineate the uneven geographies of care access and provision from a situated and embodied perspective. This approach requires not only going beyond states and markets, but assuming a non-heterocentrist position that can disrupt cisgender suppositions (Muñoz 2010; Esguerra Muelle and Quintana Martínez 2018; Esguerra Muelle 2018a). Methodologically, this implies constantly questioning the assumption that all participants are cisgender and heterosexual in order to refuse representational practices that assume the dominant cishetero-normative order as natural or default. This approach also brings to the centre the processes through which care work results from and at the same time produces social mandates of cishetero-normativity. Our feminist analysis of the urban thus involves an intersectional approach that takes seriously the place of women and other feminized subjects, spaces, and activities in the city's (re)production. We are particularly interested in pointing to the ways in which (trans)national and intra-national migrants participate in the configuration of uneven urban geographies. By analysing the experiences of paid careworkers, we highlight the importance of considering the embodied spatialities of place-making, and how they are shaped by intersectional systems of differentiation and domination (Massey 1994; Cresswell 1996; Soto Villagrán 2013).

In our collective reflections of the experience of women and other feminized migrants in cities in Colombia and Spain, we also took a descolonial approach (Rivera Cusicanqui 2014; Esguerra Muelle 2018a). Following Silvia Rivera Cusicanqui's (2014) work on the colonial matrix and the multiple ways in which the history of colonization shapes life in Latin America, the Caribbean, and Spain, we explore the complex ways in which migrants participate in the production of urban spaces. The postcolonial critique advanced by Homi Bhabha, Edward Said, and Gayatri Spivak in the 1990s focused on the relationship between European colonies of the 19th century and the Western metropolis. As a response, the decolonial project of *modernidad/colonialidad* (modernity/coloniality) proposed by Santiago Castro-Gómez, Nelson Maldonado-Torres, and Aníbal Quijano, amongst others, advanced an analysis of colonial operations in Latin America and the Caribbean since the 15th century. In contrast, the descolonial approach by Silvia Rivera Cusicanqui and other Latin American and Caribbean feminist thinkers criticizes the lack of reflexivity in relation to the conditions of production of the decolonial project embodied by modernidad/colonialidad. In discussing how the decolonial critique developed from Anglo-American academic institutions without links to anti-colonial movements, Rivera Cusicanqui says: 'Thus we have cooptation and mimesis, the selective incorporation of ideas and selective approval of those that better nourish a fashionable, depoliticized, and comfortable multiculturalism that allows one to accumulate exotic masks in one's living room...' (Rivera Cusicanqui 2012, p. 104). She also notes the problematic disregarding of the coproduction of colonial power with other forms of domination, particularly gender, in relation to a clear 'macho logocentrism' (Rivera Cusicanqui 2012, p. 106).

A descolonial approach thus helps us examine the ways in which transnational and intra-national migration constitutes a moment in which colonial relationships are constantly re-enacted. As the Colombian case shows, processes of internal and external colonialism, and their gendered, sexualized, and racialized formations, are played out through the economic, social, political, and cultural relations that provoke displacement, migration, uprooting, and exile. As the next two sections show, certain practices of urban production are the result of the ways in which (trans) national networks of care renew everyday forms of colonial domination by reinforcing the place that gendered, sexualized, and racialized bodies occupy in cities like Bogotá and Barcelona.

War, Migration, and Care: Colombian Care Workers in Spain

Since the 1990s, amidst the intensification of violence and economic crisis, migration in Colombia has grown exponentially. Currently, more than 5 million Colombians have left the country, revealing a wide array of motives of migration, socio-economic class positions, level of studies, and legal status (Echeverri Buriticá 2014). By 2016, for example, over 340 000 Colombians were living as refugees or in similar situations abroad (International Organization for Migration 2018, pp. 78–81). According to official numbers, the main countries of destination have been the United States, Venezuela, and Spain. With the Venezuelan crisis, prominent destinations are the United States (21% of migrants) and Spain (18% of migrants).[3] While migration studies have highlighted how global circuits of capital incorporate a growing number of marginalized people (due to their sexuality, gender identity, race, or national origin) into a situation of precariousness and vulnerability, we address how the concatenated processes of neo- and internal colonialism are fuelled by regimes of terror, totalitarianism, and war.

The Colombian armed conflict has long been understood as a cuase of migration, nevertheless, a feminist and descolonial view enables a deeper understanding of the entanglements of the urban with war, care, and colonialism, via the configuration of (trans)national care networks and migrant careworkers and their place in the (re)production of Madrid and Barcelona. Social and armed conflict in Colombia has resulted in high rates of forced displacement and the lack of economic opportunities, particularly for women and LGBT people. Colombian (trans)national migrants include these displaced, uprooted, and exiled women and other feminized, sexualized, and racialized subjects, articulated as links in (trans)national care networks, whose migratory experiences speak to the different forms of violence to which they have been subjected. From a descolonial approach, the continuities of strictly marked gender roles, the criminalization of certain sexual practices, and the rigidity of racial hierarchies are all forms of the internal colonialism that Rivera Cusicanqui (2012, 2014) understands as constituting the colonial matrix of domination. As we show below, these forms of domination and violence are tightly interlocked with colonial relations that sustain migrants' paid care work and their ways of inhabiting and producing the city.[4]

While broadly understood as a civil internal conflict, Colombia's war involves legal and illegal armed actors who, through an intricate network of alliances and confrontations, have made the civilian population their main target, either deliberately or by catching them in the crossfire of armed confrontations. Despite the centrality of land and capital to the dynamics of armed conflict, a descolonial take also elucidates the crucial role that geographies of colonial difference and racialization have played in processes of domination, dispossession, and death in certain regions of the country and against racialized populations (Roldán 1998; Vergara-Figueroa 2017). Moreover, there is a growing body of work which analyses gender relations, not as subsidiary, but as central to the workings of state sanctioned violence (Ojeda 2013; Viveros Vigoya 2013), sexualized forms of violence (Centro Nacional de Memoria Histórica 2017; Sánchez Parra 2018), and the multiple conflations of patriarchal violence exerted by families, communities, and legal and illegal actors (Bello 2018; Serrano 2018). Moreover, such an approach allows us to destabilize the boundaries between voluntary and forced migration, as well as to recognize the long colonial roots of war in Colombia evidenced in the central roles of race, gender, and sexuality in its functioning.

The entanglements between war, displacement, and care work provision in Madrid and Barcelona became evident in our research. Virtually all (about 97%) of the 135 women and LGBT people Esguerra Muelle interviewed are victims of the internal armed conflict. This was the case of Yacky, a woman working at a cleaning company in Barcelona, who reflected on her own migratory trajectory: 'I'm from a little town [near Cali], we had to leave when I was little, we arrived to Cali and I had, well, a tough life. I don't have any nice memory from Colombia, you know. (…) [T]o say the truth, I've had a very difficult life and coming here wasn't also easy (…)' (personal interview, Barcelona, May 2018).[5] In her accounts, Yacky evidenced the connections between structural violence, the social armed conflict, internal displacement, and international migration. As migration fodder, Yacky entered (trans)national networks of care both in Colombia and Spain.

These connections were also evident in the experience of migrants who did not entirely comply with cisgender and heterosexual norms. All the trans people and lesbians who participated in the research associated their decision to migrate with their gender identities and sexual orientation. Their displacement was the result of intertwined forms of lesbophobia and transphobia which rendered their existences non-viable. This is the case of Wendy, a trans woman with a physical disability, who, at the time of our interview, was a domestic worker in Bogotá and was only paid in kind for her work: 'From Chaparral [a rural town in Tolima] I migrated to Ibagué [a nearby city] and then I came to Bogotá to work. There, there were no opportunities to work or study (…) I couldn't dress up, wear makeup or have friends, nothing. There weren't trans women, just a few (…) there it wasn't possible because of the guerrillas' (personal interview, Bogotá, August 2018).

A similar experience was reflected by Camilo, a self-identified transexual man, who migrated from Cali to the US, the Netherlands, and then to Spain. He lived in Barcelona and was a domestic worker at the time of our interview:

> Why someone [like me] who wasn't afraid to cross through *el hueco* [the Mexico-US border] is afraid to lose his job? Because the situation here is really hard (...). I haven't changed my documents, so I look for jobs as a man, but my name is female on paper, very few people understand or accept that (...). All work I've done since I left Colombia is (...) very hard work [like] (...) cleaning rooms and making beds at a hotel (...) and painting that hotel. I left for different reasons, *a mí me mandaron a matar* [they sent someone to kill me] because I had a relationship with the Mayor's niece (...) it was platonic, but her family found out. (personal interview, Barcelona, May 2018)

While their stories could be read as those of seeking a more inclusive and less violent place in cities in the global North, Wendy's and Camilo's experiences of working and living in Madrid and Barcelona tell a more complex story. Their positionality assigns them to a place in the occupational hierarchy where they provide long hours of devalued and underpaid work, at the same time reproducing their subjectivities and naturalizing their embodied disposition and capacity to carry out care work. While detaching them from the specific contexts they were forced to leave, gendered, sexualized, and racialized forms of discrimination and exclusion still define how they inhabit the city. The demand for feminized care work from cities in Spain, mostly but not limited to cleaning services and childcare, cannot be understood outside the historical, localized, and embodied configurations of a gendered, sexualized, and racial order that is enacted and spatialized on an everyday basis. This is evidenced, for example, in the experiences of racism and xenophobia that participants reported in relation to their national origin and their representation as 'sudacas', a derogatory term used to signal a Latin American origin, and the way this shaped their itineraries, detours, and meeting places.

The continuity of colonial domination naturalizes the idea that feminized bodies, and in particular those of racialized women, are born to care for and to serve others.[6] Labour practices in both Colombia and Spain do not usually provide these workers with equitable remuneration, employment insurance, healthcare, the right to rest, or to a schedule. Such devaluation of their work translates into a continuum of forms of violence that vary from sexual and psychological violence, to xenophobia, misogyny, transphobia, or lesbophobia, and economic violence, as expressed in trafficking, servitude, and slavery. Camilo said when describing his job: 'My salary is shit (...) I'm supposed to work [and get paid for] four hours, but it's usually five and a half, more or less. (...) Every day, because four hours are never enough for all I have to do (...)' (personal interview, Barcelona, May 2018). In addition to precarious working conditions, the great majority of participants felt subjected to discriminatory discourses regarding their servility, as well as noting the persistence of discursive constructions of gendered and racial inferiority, the hypersexualization of non-white women, and the naturalization of care work as their responsibility (personal interviews, 2018). These dynamics, along with constant fear of being arrested and deported for the case of international migrants, are recurrent elements of their experiences in/of Madrid and Barcelona.

For example, Alana was displaced by internal violence, in and out of the war, firstly from Buenaventura and later from Cali. She decided then to migrate to Barcelona, leaving her son behind. Months later, when her son arrived with his father, they were deported. Alana explains this situation in terms of blatant racism and connects it with her own experience in relation to her work:

> (...) that seems to me like an injustice because there was no way to return them because they had everything in order (...) everything a tourist has to bring (...) Going as a tourist to a country is not a crime, you know? And more so if you bring everything in order. That's why I tell you it was more that the people who stopped them are racists. They did not like them, because they were black, and they were colored people and they were migrants. They locked them in a cell, my son almost died because he was locked up with a lot of people, the child got sick (...). (...) There are racist people here. I also work in a place where the supervisor was a Catalonian and she was very racist (...) she treated the girls very badly and you see that, she does not like dark people. Once she treated me so badly (...) she told me that I was blacker than a toilet. I had never been mistreated like this (...) nor in my own country, and here that happens to me. (personal interview, Barcelona, July 2018)

Alana's experience of racism also reveals colonial forms of hypersexualizing black bodies: 'I went to an employment agency where they told me they would help me to get a job, no matter if I had no documentation. (...) But no task ever came [she was not assigned any jobs], except for men proposing me indecent things (...) In fact, I had to tell one of them that if he continued harassing me, I was going to throw him out to the police, and that's how he left me a bit alone' (personal interview, Barcelona, July 2018)

Taking into account the continuum of violence expressed through their experiences is central to understanding the many ways in which immigrant careworkers inhabit and remake the city. Many of them, in response to these intertwined forms of violence, manifested a feeling of entrapment in their workplaces. Mostly domestic workers, their life passes between their residences and the homes where they work. Most of them undocumented, insisted on their fear of being deported. Subjected to racism and xenophobia, their experiences of the city were profoundly marked by a limited mobility. Diana's testimony is a clear example of this. In relation to her live-in working situation, she said:

> My situation as migrant… is so hard. (…) You must be very careful with police raids at the subway entrances (…) you should not be around *asking for it*. So you have always to be like shut up in your work place, because you cannot afford having to make that sacrifice [referring to migrating and working in precarious conditions], so that overnight you get caught for something so simple as using the subway and being deported. (personal interview, Madrid, September 2017).

The experiences detailed in the quote above resonate with Amy Ritterbusch's work (2016) in Bogotá, which found that girls, youth, and cis and trans women who are sex workers have very limited possibilities to participate in urban life. Most of them have been internally displaced at some point of their lives, hence their lives are profoundly shaped by police terror and patterns of urban socio-spatial segregation. Along with Ritterbusch and other Latin American urban studies (Massolo 2004; Jirón Martínez 2007; Silva 2009; Soto Villagrán 2018), our research highlights the role of gender and sexuality in the production of urban space, often involving dynamics of violence and exclusion. In the next section we turn to Bogotá, in order to understand a different care regime and its articulations with violence, migration, and the ways in which care work contributes to shaping urban space.

Communitarian Mothers in Colombia

In Colombia, 'communitarian mothers' are women responsible for providing early childhood care for children whose parents cannot provide for them, generally because of their precarious economic situation. The state formalized the concept of communitarian mothers in 1986, and currently there are around 69 000 women providing these services across the country, most from rural and working-class backgrounds, and some of them Afro-Colombian and Indigenous. While taking care of young children in their homes and providing a varied set of care services, communitarian mothers' work is only recognized officially as volunteer work. Between 1986 and 2014, the Colombian government paid them no more than half the legal minimum wage (that was about USD200 per month in 2014) on the assumption that this supposedly volunteer work is simply an extension of women's role in society.

In 2016, the Colombian Constitutional Court recognized that the communitarian mother model of care resulted in labour exploitation in terms of gender-based structural violence against communitarian mothers. Nonetheless, in 2018, the Constitutional Court concluded that, due to the voluntary nature of their work, they were not eligible for pensions: according to a press release 'communitarian mothers do not have the right to a pension to dignify their old age, and are condemned to indigence' (Esguerra Muelle 2018b). In August 2018, communitarian mothers decided to undertake a hunger strike, seeking to claim their pensions and recognition as workers. After eight days, the strike dissolved because of many of the women's advanced age and poor health. To date, the Constitutional Court has not decided on the nature of the link between communitarian women and the state; it has only ruled on the government objection to the bills that provided for the possibility of direct contracting the communitarian mothers by the ICBF (Colombian Institute for Family Wellbeing) (Sentencia C110 2019). Thanks to this constitutional judgement, in recent days (7 July 2019), after more than 30 years of workers' continuous social struggles, a Congress bill is being processed in order to establish the guidelines on the employment relationship of communitarian mothers and the state. As in the case of Colombian careworkers who migrated to Spain, we found that most communitarian mothers are forcedly displaced women, coming from rural areas affected by war. They provide care both to children of internally displaced women and to children of transnational migrants, who in turn are mostly caregivers in other countries.

This is the case of Guillermina, a displaced communitarian mother in Bogotá: 'When I came here to Bogotá [in 1974], I was 14, it was when my mom let me come to take care of my sister's kids (…). I took care of my two nieces and another three children in a space with one room and a kitchen. It was always difficult, because I came from the countryside and it was like being oppressed (…)' (personal interview, Bogotá, March 2018). After living in Bogotá for a year and a half, Guillermina returned to her rural town and started to work at the Mayor's home. Tired of his unsolicited sexual advances, she returned to her parent's home just before the guerrillas displaced her and her family. She remembers how the presence of illegal actors, including guerrillas, became constant in her town. 'I used to go to see my parents, but one had to encounter them (…). In 2008

I remember I came with my son and they [FARC guerrillas] put a gun to him and asked him to join them (...)' (personal interview, Bogotá, March 2018). Guillermina returned to Bogotá and after working for several years cleaning homes she became a communitarian mother.

The experiences of war and forced displacement of research participants make evident the tight relations between Colombia's social and armed conflict and the configuration of (trans)national networks of care. Many of the participants reflect on how they had to leave their rural hometowns to migrate on several occasions. Becoming migration fodder, they end up participating in gendered and racialized regimes of care saturated with precarity, exclusion, and the continuum of intertwined forms of violence. Research participants often referred to exploitative working conditions. Communitarian mothers' workdays extend beyond 15 hours, as they must alternate between the care of their own homes with their work as careworkers for others. They also expressed the number of risks they face at work and how, in order to make ends meet, they have had to undertake other care work in environments where they are mistreated and further exploited.

Teresa, a woman displaced from Cali, stated that none of her *compañeras* had insurance for risks encountered in the workplace. In her case, she also did not have a pension. The half minimum wage they were supposed to receive was also very difficult to claim. She stated: 'Between 1997 and 2000 we had many problems with that payment ..., they took up to six months to pay us ... so debts accumulated and then after six months one was left outside the [social security] system ... so one had to enter again and that implied a lot of time and lines and travelling to Cali...' (personal interview, Cali, February 2018).

In their recounting, communitarian mothers made evident the ways in which they carry out their work at the expense of their own bodies and existences. According to Cecilia: 'I haven't been feeling well. I've visited the doctor because I have some trouble with my spine (...), so the doctor tells me "Ceci, you need to change your routine." So, I tell him about my work, and how I don't stop and move from here to there (...). The doctor says [the pain] is because of my work' (personal interview, Bogotá, April 2018).

As in the case of careworkers in Spain, the life trajectories of communitarian mothers reinforce a colonial hierarchy of difference that assigns certain – impoverished, racialized, gendered, and sexualized – bodies to a place in society where they are expected or forced to work as domestic workers, nurse's aides, hotel maids, and sex workers. This confirms that the configuration of care regimes in relation to forced displacement reinforce an exploitative social order, which re-enacts colonial forms of domination.

The case of communitarian mothers relates to (trans)national networks of care as their subject positions situate them as migration fodder. The same gendered, sexualized, and racialized bodies end up as careworkers either in Colombian cities or overseas. Moreover, communitarian mothers often take care of the children of Venezuelan migrants, Colombian returnees, and other internally displaced people. And while they are not state employees, communitarian mothers are, for example, the ones in charge of providing care to malnourished children of preschool age coming from Venezuela (Consejo Nacional de Política Económica y Social 2018).

Communitarian mothers' care responsibilities are not supported by any monetary recognition for the state's use of their home space and personal property. They also assume most of the costs for the provision of water and energy. At the same time, they are subjected to routine inspection and control visits through which the state restricts and regulates the use of the home space. Yolis, Amanda, Dominga, and Teresa, amongst other participants, outline the state's close supervision and how each state official made its own rules and policies. Yolis tells how she organized the space of her house when she was a communitarian mother so it would fit her work. She, like other communitarian mothers, often referred to not having social areas in their houses, because these areas are dedicated to serving the children they care for (personal interviews). They are required to make all the spatial adjustments required by the ICBF (Colombian Institute for Family Wellbeing) with their own resources. As Amanda tells us: 'I applied for a loan as a communitarian mother at that time. And then, with that, *mija* [dear], I put myself to the task of finishing the back of the house, and I started to do the bath (…)' (personal interview, Cali, May 2018). Yolis complained of how the ICBF deducted monthly the loan from the compensation (half minimum wage) she received. This is corroborated by Dominga: 'If the communitarian mother is going to buy a bag of cement, she has to buy it with her effort. And the ICBF requests the house, the living room, everything, they demand everything from you and they do not give anything back' (Personal Interview, Cartagena April 2018). Teresa also says that 'the ICBF demands exclusivity (…) it is as if we belong to them [state officials]. Yours!? Well, I don't feel like I belong to anyone' (personal interview, Cali, February 2018). Cecilia also pointed to the difficulties that result from the state's supervision of her work, still deemed as voluntary: 'sometimes one has to leave the children stay still, run to the computer and download the form they sent (…) or right now one has to organize and file 14 folders (…). When it's not the characterization file it's the valorization file (…), it's never eight hours of work' (personal interview, Bogotá, April 2018).

Communitarian mothers' care work has a role to play in shaping urban space. Their homes are places of production and social reproduction, closely subjected to state supervision. At the same time, they are important community spaces where babies and children spend most of their time, as well as a key point of reference for neighbours. In many cases, their homes are also the place where they carry out activities such as hairdressing and selling cosmetics. As in the case of careworkers in Madrid and Barcelona, their practices are fundamental for the production of the city. Their work is what sustains life and provides the necessary conditions for other spheres of work to happen. They too participate in the cities they live in through restricted mobilities, limited access to urban resources, and relational forms of violence.

Conclusion

Through the analysis of the experiences of careworkers who migrated to urban contexts in Colombia and Spain, this chapter has examined the connections between social reproduction, urban space, war, forced displacement, and the configuration of (trans)national networks of care. From a feminist descolonial standpoint, our research shows how in and through these tramas – as fabric, narration, and plot – gendered and racialized forms of colonial domination are re-enacted and reproduced. Three important points in relation to the production of urban spaces emerge from this analysis: the centrality of relational forms of violence and care; the need to attend to intersectional embodied spatialities; and the relevance of the geographies of home and other domestic spaces of care work as key sites of urban production.

Throughout the chapter we have highlighted the precarious conditions of careworkers and the ways in which their embodied spatialities are highly constrained. Many migrants and displaced people, particularly feminized subjects, end up working in different forms of care work. Whether part of a strategic choice or not, the isolation of working in domestic spaces undermines workers' health and existence and renders isolation as yet another form of poverty. Many migrants cannot leave their places of work due to their contracts or fear of migratory raids, or because they are unfamiliar with the urban space in which they live, revealing the ways in which their exploitation, subordination, and criminalization fit within a neoliberal, hyper-vigilant, and revanchist city. In the case of communitarian mothers, their work is performed at home, under conditions they can seldom control. Through their everyday itineraries and work, careworkers in Spain and in Colombia play a fundamental role in the production of urban geographies, which are highly uneven. Careworkers traverse large distances when evicted from their hometowns but inhabit the cities they arrive to with a de facto confinement sentence that limits them to domestic spaces and pendular trajectories. As feminized subjects, racially marked and from rural origins, careworkers are subjected to a continuum of violence. Their engagement with processes of urban place-making, while limited by this continuum, is key for understanding the (re)production of urban life, as well as the contested ways in which urban space is constantly configured.

Care work and careworkers in cities like Bogota, Cartagena, Madrid, and Barcelona are key for better understanding contested processes of urban production. These cities do not just absorb certain bodies, but are actively produced by them. While being outsiders, careworkers are central to the production of the urban. Their work sustains the symbolic and material maintenance of life. Also, responding to their experiences of marginalization, research participants were often involved in community networks and many of them end up organizing against issues such as social discrimination and gentrification, as evidenced in the Tancada Migrante (Migrant Lockup) carried out by migrant workers in Barcelona in May to August 2018 or in the hunger strike performed by communitarian mothers in Bogota in August 2018.

By analysing migrant careworkers' experience, we point to the importance of considering the relations between forced displacement, intra-national and international migration, and (trans)national care networks in relation to urban space. We argue that the embodied spatialities of women and other feminized subjects make evident the need for more feminist descolonial studies of urban production.

Acknowledgements

This research was made possible by funding from CIDER and Vicerrectoría de Investigación at Universidad de los Andes, and the support of the Center of Occupational Health CiSAL of Universidad Pompeu Fabra. We thank Ivette Sepúlveda, Laura Castrillón, Alí Majul, María de Los Ángeles Balaguera, Eliza Enache, Juan Manuel Guerrero, José Antonio de la Cruz, and Carlos Huertas for their research assistance. We would also like to thank Linda Peake, Elsa Koleth, and darren patrick/dp for their comments and suggestions, as they truly constituted an opportunity for careful dialogue and exchange.

Notes

1 According to official figures, there are 3 627 350 women victims of forced displacement, compared to 3 444 552 men; and there are 1738 LGBT individuals (UARIV 2015).

2 Esguerra Muelle used snowball sampling to employ in-depth biographical interviews, individual and collective social cartographies, discussion groups, and participant observation in domestic spaces. In total, she interviewed 135 people in 7 cities: 23 in Madrid, 16 in Barcelona, 23 in Bogotá, 25 in Medellín, 21 in Cali, 22 in Cartagena, and 5 in Ibagué. In total 93.3% of participants were cisgender women, 3.7% transgender women, 1.5% cisgender men, and 1.5% transgender men or non-binary persons. Of the participants, 51.9% identified as heterosexual and 41.5% declared themselves 'normal', and 4.4% identified as lesbians. In total, 96.2% of interviewees were from rural origin. In terms of ethnicity, 18.5% identified as mestizx, 18.5% identified as Afro-Colombian, 7.4% as Black, 6.8% as Caribbean, 6.7% as Indigenous, and 1.5% as Afroindigenous. In terms of age, 43% were between 27- and 40-years old, 31.9% between 40- and 57-years old and 16.3% above 57 years of age. Of all the participants, 31.1% were transnational migrants and 78.5% were internal migrants and 62.2% had been internally displaced due to the armed conflict. Ojeda's and Fleischer's research is also based on snowball sampling to employ biographical interviews and social cartographies.

3 In 2018, 935 593 Venezuelans were living in Colombia, a number that increased to 1 174 743 by February 2019 (Migración Colombia 2019). Colombia is also an important site of transit migration of people from Asia, Africa, and the Caribbean (mainly Cuba and Haiti), who seek to cross Central America to reach the United States (Heinrich Böll Stiftung 2017).

4 There is also an important point to be made in relation to the direct participation of large multinational corporations in Colombia's war (as in the case of Chiquita/Dole and Pacific Rubiales), and the benefits to them resulting from the elimination of dissident rural communities who oppose their megaprojects. It is also important to connect foreign direct investment with the the deepening of inequalities in the country. For example, in 1995, the Spanish company, Aguas de Barcelona, began to provide water and sanitation services to the city of Cartagena. A decade later, a corruption scandal broke out as the company had failed to fulfill its obligation to provide water and sanitation services to impoverished neighbourhoods (at least a third of the city's population) and had caused serious ecological and health damage (Ramiro, González, and Pulido 2007).

5 Translations from Spanish by the authors.

6 In 1999, 25.2% of immigrant women in Spain were from Latin America; this increased to 44.9% by 2007 (Instituto Nacional de Estadística 2017). According to the National Ministry of Work, Migration and Social Security

(2009), during 2008, Spain issued more than 790 000 work permits to immigrants, of which 54% were issued to Latin American immigrants, about half of these to Ecuadorean, Colombian, Peruvian, and Venezuelan individuals. Of these work permits, nearly half were issued to domestic workers.

References

Arango, L.G. (1993). Religion, family and industry in the transmission of values. In: *Between Generations. Family, Models Myths And Memories* (eds. D. Bertaux and P. Thompson), 51–68. London: Oxford University Press.

Arango, L.G. (ed) (2018). *Género y cuidado teorías, escenarios y políticas.* Bogotá: Editorial Universidad Nacional de Colombia; Editorial Javeriana; Editorial Universidad de los Andes. 270. ISBN: 978-958-781-221-3

Bakker, I. and Silvey, R. (2008). *Beyond States and Markets: The Challenges of Social Reproduction.* London: Routledge.

Bello, A. (2018). *Un carnaval de resistencia: Memorias del reinado trans del río Tuluní.* Bogotá: Centro Nacional de Memoria Histórica.

Bettio, F., Simonazzi, A. and Villa, P. (2006). Change in care regimes and female migration: The 'care drain' in the Mediterranean. *Journal of European Social Policy* 16 (3): 271–285.

Carbado, D.W., Crenshaw, K., Williams Mays, V.M et al. (2013). Intersectionality. *Du Bois Review: Social Science Research on Race* 10 (2): 303–312.

Centro Nacional de Memoria Histórica (2017). *La guerra inscrita en el cuerpo. Informe nacional de violencia sexual en el conflicto armado.* Bogotá, DC: Centro Nacional de Memoria Histórica.

Consejo Nacional de Política Económica y Social (2018). Documento CONPES 3059. *Departamento Nacional de Planeación – Ministerio de Hacienda y Crédito Público. República de Colombia.* https://colaboracion.dnp.gov.co/CDT/Conpes/Económicos/3959.pdf (accessed 6 May 2019)

Cresswell, T. (1996). *In Place–Out of Place: Geography, Ideology, and Transgression.* Minneapolis, MN: University of Minnesota Press.

Dorlin, E. and Bidet-Mordrel, A. (2009). *Sexe, race, classe: pour une épistémologie de la domination. Actuel Marx confrontation.* Paris: Presses universitaires de France.

Echeverri Buriticá, M.M. (2014) A los dos lados del Atlántico. Reconfiguraciones de los proyectos migratorios y la vida familiar transnacional de la población colombiana en España. *Papeles Del CEIC* 2: 1–28. doi:10.1387/pceic.12988

Esguerra Muelle, C. (2014) Dislocación y borderland: Una mirada oblicua desde el feminismo descolonial al entramado migración, régimen heterosexual, (pos)colonialidad y globalización. *Universitas Humanistica* 78: 137–161. doi:10.11144/Javeriana.UH78.dbmo

Esguerra Muelle, C. (2018a). Coloniality, colonialism an decoloniality: Gender, sexuality, and migration. In: *The Routledge Handbook of Latin*

American Development (eds. J. Cupples, M. Palomino-Schalscha, and M. Prieto), 54–63. London: Routledge.

Esguerra Muelle, C. (2018b). Ni madres ni voluntarias. No hay paz sin cuidado. *El Espectador.* https://colombia2020.elespectador.com/opinion/ni-madres-ni-voluntarias-no-hay-paz-sin-cuidado (accessed 6 May 2019)

Esguerra Muelle, C. (2019). Etnografía, acción feminista y cuidado: Una reflexión personal mínima. *Antípoda. Revista de Antropología y Arqueología* 13 (35): 91–111. doi:10.7440/antipoda35.2019.05

Esguerra Muelle, C. and Quintana Martínez, A. (2018). 'Tu vida también es mi país': Sexualidades disonantes y fugas de género en Liliana Felipe y Jesusa Rodríguez. *Cuadernos de Musica, Artes Visuales y Artes Escenicas* 13 (1): 61–84. doi:10.11144/javeriana.mavae13-1.tvte

Esguerra Muelle, C., Sepúlveda Sanabria, I., and Fleischer, F. (2018). *Se nos va el cuidado, se nos va la vida: Migración, destierro, desplazamiento y cuidado en Colombia (No. 3).* Bogotá. www.cider.uniandes.edu.co/Documents/Publicaciones/Senosvaelcuidado_senosvalavida.pdf (accessed 6 May 2019)

Fleischer, F. and Marín, K. (2019). Atravesando la ciudad. La movilidad y experiencia subjetiva del espacio por las empleadas domésticas en Bogotá. *EURE* 45 (135): 27–47. doi:10.4067/S0250-71612019000200027

Heinrich Böll Stiftung (2017). *Ir, venir, quedarse, seguirle – Facetas de la migración en América Latina.* www.boell.de/en/2018/08/17/annual-report-2017 (accessed 6 May 2019).

Hochschild, A.R. (2000). Global care chains and emotional surplus value. In: *On the Edge: Living with Global Capitalism* (eds. A. Giddens and W. Hutton), 130–146. London: Jonathan Cape.

International Organization for Migration (IOM) (2018). *World Migration Report 2018.* www.iom.int/wmr/world-migration-report-2018 (accessed 6 May 2019)

Instituto Nacional de Estadística (INE) (2017). *España en Cifras.* www.ine.es/prodyser/espa_cifras/2015/files/assets/common/downloads/publication.pdf (accessed 6 May 2019).

Jirón Martínez, P. (2007). Implicancias de género en las experiencias de movilidad cotidiana urbana en Santiago de Chile. *Revista Venezolana de Estudios de La Mujer* 12 (29): 173–197.

Klammer, U. and Letablier, M.-T. (2007). Family policies in Germany and France: The role of enterprises and social partners. *Social Policy and Administration* 41 (6): 672–692.

Leonard, S. and Fraser, N. (2016). Capitalism's crisis of care. *Dissent* (Fall). www.dissentmagazine.org/article/nancy-fraser-interview-capitalism-crisis-of-care (accessed 6 May 2019).

Marcus, G. (1995). Ethnography in/of the world system: The emergence of multi-sited ethnography. *Annual Review of Anthropology* 24: 95–117.

Marcus, G. (2018). The contemporary desire for ethnography and its implicaion for anthropology. In: *Transforming Ethnographic Knowledge* (eds. K.M. Clarke and R. Hardin), 73–90. Madison, WI: University of Wisconsin Press.

Massey, D. (1994). *Place, Space and Gender*. Minneapolis, MN: University of Minnesota Press.

Massolo, A. (2004). *Una mirada de género a la Ciudad de México*. México, DF: UAM Azcapotzalco, Red Nacional de Investigación Urbana (RNIU).

Meehan, K. and Strauss, K. eds. (2015). *Precarious Worlds: Contested Geographies of Social Reproduction*. Athens, GA: University of Georgia Press.

Migración Colombia (2019). *Boletín anual de estadísticas*. Bogotá, DC. www.migracioncolombia.gov.co/phocadownload/BoletínEstadísticoFlujosMigratorios2018_032019.pdf (accessed 6 May 2019)

Mitchell, K., Marston, S., and Katz, C. (2004). *Life's Work: Geographies of Social Reproduction*. London: Blackwell.

Muñoz, L. (2010). Brown, queer and gendered: Queering the latina/o 'streetscapes'. In: *Queer Methods and Methodologies* (eds. K. Browne and C. Nash), 55–67. London: Routledge.

National Ministry of Work Migration and Social Security (2009). *Estadística de Autorizaciones de Trabajo a Extranjeros*. www.mitramiss.gob.es/estadisticas/Pte/welcome.htm (accessed 6 May 2019)

Ojeda, D. (2013). War and tourism: The banal geographies of security in Colombia's 'retaking'. *Geopolitics* 18 (4): 759–778.

Ojeda, D. and González, C. (2018). Elusive space: Peasants and resource politics in the Colombian Caribbean. In: *Land Rights, Biodiversity Conservation and Justice: Rethinking Parks and People* (eds. S. Mollett and T. Kepe), 88–106. New York: Routledge.

Parreñas, R.S. (2012). The reproductive labour of migrant workers. *Global Networks A Journal of Transnational Affairs* 12 (2): 269–275.

Ramiro, P., González, E. and Pulido, A. (2007). Las Multinacionales Españolas en Colombia. *Observatorio de Multinacionales en América Latina*. http://omal.info/spip.php?article737 (accessed 6 May 2019).

Ritterbusch, A.E. (2016). Mobilities at gunpoint: The geographies of (im)mobility of transgender sex workers in Colombia. *Annals of the American Association of Geographers* 106 (2): 422–433.

Rivera Cusicanqui, S. (2012). Ch'ixinakax utxiwa: A reflection on the practices and discourses of decolonization. *South Atlantic Quarterly* 111 (1): 95–109.

Rivera Cusicanqui, S. (2014). La noción de 'derecho' o las paradojas de la modernidad postcolonial: Indígenas y mujeres en Bolivia. In: *Tejiendo de otro modo: feminismo, epistemología y apuestas descoloniales en Abya Yala* (eds. Y.E. Miñoso, D.G. Correal, and K.O. Muñoz), 121–134. Popayán, CO: Editorial Universidad del Cauca.

Roldán, M. (1998). Violencia, colonización y la geografía de la diferencia cultural en Colombia. *Análisis político* 35: 3–26.

Sánchez Parra, T. (2018). The hollow shell: Children born of war and the realities of the armed conflict in Colombia. *International Journal of Transitional Justice* 12 (1): 45–63.

Serrano, J.F. (2018). *Homophobic Violence in Armed Conflict and Political Transition*. London: Palgrave-McMillan.

Silva, J.M. (2009). *Geografias Subversivas: discursos sobre espaço, gênero e sexualidades.* Ponta Grossa: Todapalavra.

Soto Villagrán, P. (2013). Repensar las prácticas espaciales: rupturas y continuidades en la experiencia cotidiana de mujeres urbanas de la Ciudad de México. *Revista Latinoamericana de Geografía e Genero* 4 (2): 2–12.

Soto Villagrán, P. (2018). Hacia la construcción de unas geografías de género de la ciudad. Formas plurales de habitar y significar los espacios urbanos en Latinoamérica. *Perspectiva Geográfica* 23 (2): 13–31.

Tronto, J.C. (2013). *Caring Democracy : Markets, Equality, and Justice.* New York: NYU Press.

Tronto, J.C. (2016). *Who Cares?: How to Reshape a Democratic Politics.* Ithaca, NY: Cornell University Press.

UARIV Unidad de Atención y Reparación Integral de Víctimas (2015). *Enfoque de orientaciones sexuales e identidades de género.* www.goo.gl/shaqLj (accessed 6 May 2019).

UARIV Unidad de Atención y Reparación Integral de Víctimas (2019). Registro Único de Víctimas, *Red Nacional de Información. Información al servicio de las víctimas.* www.rni.unidadvictimas.gov.co/RUVcorte2018 (accessed 6 May 2019).

Ungerson, C. (1999). Personal assistans and disabled people: An examination of a hybrid form of work and care. *Work, Employment & Society* 13 (4): 583–600.

Ungerson, C. (2004). Whose empowerment and independence? A cross-national perspective on 'cash for care' schemes. *Ageing and Society* 24 (2): 189–212.

UNHCR (2017). *Global trends: Forced displacement in 2017.* https://www.unhcr.org/5b27be547.pdf (accessed 6 May 2019).

Vergara-Figueroa, A. (2017). *Afrodescendant Resistance to Deracination in Colombia: Massacre at Bellavista-Bojayá-Chocó.* First Edition. Cham: Palgrave Macmillan.

Viveros Vigoya, M. (2013). Género, raza y nación. Los réditos políticos de la masculinidad blanca en Colombia. *Maguaré* 27 (1): 71–104.

Viveros Vigoya, M. (2016). La interseccionalidad: una aproximación situada a la dominación. *Debate Feminista* 52: 1–17.

10

Tenga Nehungwaru

Navigating Gendered Food Precarity in Three African Secondary Urban Settlements

Belinda Dodson (Department of Geography, University of Western Ontario)
Liam Riley (Balsillie School of International Affairs, Wilfrid Laurier University)

Introduction

While we were writing this chapter in 2019, the army and police in Zimbabwe were destroying vendors' stalls on city streets. This resembled an earlier crackdown in 2005 known as Operation *Murambatsvina*. In both cases, clearances were presented as an attempt to clean up the city by removing vendors from unauthorized spaces. Streetnet (2019), an international alliance of street vendors, immediately issued a statement of solidarity, decrying the effect on Zimbabweans' livelihoods: 'We stand in full solidarity with all street vendors and informal traders of Zimbabwe, and call for an end to this operation – an attack on the livelihoods of the working poor which the world is watching.'

Such evictions are an attack on social reproduction. In Zimbabwe, as in other countries in the region, the informal sector is essential to the lives and livelihoods of the urban poor. Many adults and children purchase and sell food through various forms of informal food retail: street vendors; doorstep and roadside stalls; neighbourhood kiosks; and larger informal

A Feminist Urban Theory for our Time: Rethinking Social Reproduction and the Urban,
First Edition. Edited by Linda Peake, Elsa Koleth, Gökbörü Sarp Tanyildiz, Rajyashree N. Reddy & darren patrick/dp.
© 2021 John Wiley & Sons Ltd. Published 2021 by John Wiley & Sons Ltd.

markets, often at transportation hubs or other sites where people congregate. There are also formal food markets, such as those run by municipal authorities who lease stalls to vendors, and formal food retailers, including increasing market penetration by supermarket chains (Crush and Frayne 2011; Battersby and Watson 2018). Yet, African city authorities' attitudes to informal food trade are frequently ambivalent and occasionally hostile (Tacoli 2017; Smit 2018). This has uneven and gendered effects on urban residents. Examining how people experience and navigate urban food systems reveals the political, economic, and social forms and processes through which food becomes accessible – or not – to individuals, households, and communities (Riley and Dodson 2014; Tevera and Simelane 2016; Mackay 2019). Looking at cities through food and food systems can thus provide a way to make social reproduction more central in urban theory, policy, and planning.

There is also growing recognition of the need to bring cities more centrally into discussions of food security (Ruel and Garrett 1999; Tacoli 2017). From either approach, gender and gender relations are shown to be fundamental to understanding urban food access and food systems. First, gender affects individual and household food security (Levin et al. 1999; Floro and Swain 2013). Women and female-headed households are not inevitably or invariably less food secure. Rather, the connection between gender and individual or household-level food security operates through additional, intersecting factors such as household size and composition, men's and women's differential inclusion in the labour market, unequal access to financial and other assets, and social norms of gender roles and relations – all of which are altered by urbanization. Second, in what is still a strongly gendered division of household labour, women remain largely responsible for food procurement, allocation, and preparation within the domestic sphere. Their reproductive labour sustains urban households and helps mitigate household income poverty (Tacoli 2012; Chant 2014; Cock 2016). Third, women are important actors in the urban food economy as traders, processors, and producers, especially in the urban informal sector so essential to the food procurement strategies of the urban poor in the global South (Tinker 1997; Hovorka, de Zeeuw, and Njenga 2009; ILO and WIEGO 2013; King 2016). Together, as Jacklyn Cock has argued in the context of South Africa, this means that 'women's involvement in food production, procurement and preparation is a fundamental feminist issue' (2016, p. 121). Essential to family and social reproduction, women's paid and unpaid food-related labour simultaneously reproduces patriarchal family structures, limits women's participation in other social and economic opportunities, and compensates for state failure to provide adequate social protection.

This chapter presents findings from a multi-partner international project titled 'Governing Food Systems to Alleviate Poverty in Secondary Cities in

Africa' (branded Consuming Urban Poverty, or CUP). The project focused on secondary cities, broadly defined. Although much urban research in the global South has focused on national capital or large cities, significant urban population growth in Africa is occurring in intermediate- and smaller-sized cities (United Nations 2015; Battersby and Watson 2018). As CUP findings have shown, 'governance capacity issues are particularly evident in secondary cities, which are also the urban areas with the most pressing development challenges' (Battersby and Watson 2018, p. 7). CUP examined urban food systems and food security in three African secondary urban settlements: Kitwe in Zambia, Kisumu in Kenya, and Epworth in Zimbabwe (see Figure 10.1). The intent was not to claim that these are representative of secondary cities in Africa, but rather to explore each in

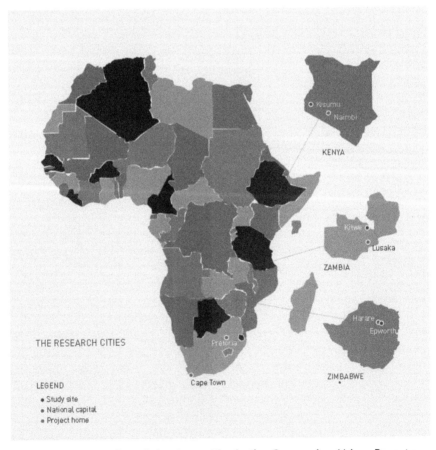

Figure 10.1 Location of the three cities in the Consuming Urban Poverty project. (Consuming Urban Poverty 2019)

context in order to gain understanding of how urban food systems and their governance influence urban food security and food poverty.

Our analysis draws primarily on data from qualitative, household-level interviews to shed light on gendered lived geographies of food access and food poverty. The title of the chapter, *Tenga Nehungwaru*, means to 'buy wisely'. In interviews conducted by CUP's Zimbabwe project team, it was used to describe how women use their knowledge, experience, and social networks to know where, when, and how to obtain food in order to stretch limited household budgets and feed their families in conditions of economic insecurity. Findings reveal how food access and food poverty are embedded in household form, social norms around gender roles and occupations, and kinship and community networks. Read in conjunction with other CUP findings, they show how governance of urban food systems can undermine the food access strategies and reproductive labour that sustain urban households and communities.

Food and Social Reproduction in African Cities

Food insecurity is strongly correlated with household income poverty in African urban economies, with high rates of unemployment and informal employment and where purchasing is the predominant form of food access (Crush, Frayne, and Pendleton 2012; Battersby and Watson 2018). Understanding food security in rapidly-urbanizing, low-income contexts therefore requires an approach that connects individual, household, neighbourhood, city, and national scales (Riley and Hovorka 2015; Tacoli 2017). Informal sources of food purchases are important for many urban Africans, despite city policies and actions that are often unsupportive or even repressive of informal trading (Young and Crush 2019). In many places, women are disproportionately engaged in informal food trade, making them particularly vulnerable to restrictions or prohibitions on certain forms or sites of food vending. The informal sector more generally is an important source of employment and livelihood, providing limited and unreliable incomes and causing vulnerability to food and other price fluctuations. Poor urban households typically spend most of their income on food, and often incur a 'poverty penalty' by paying higher prices for buying food on a day-to-day basis close to where they live. Many of the urban poor live in informal (or slum) settlements, where health risks and non-income dimensions of lived poverty exacerbate food insecurity. These conditions and constraints can fall especially hard on women, who are primarily responsible for food storage and preparation in persisting gendered patterns of household labour.

Individual and household food security also reflect intra-household processes and power relations, along with social relations beyond the family or household scale. Intra-household dynamics of bargaining power, income distribution, and labour allocation are important determinants of food security (Stevano 2019). The home remains the main site of food consumption, although street foods are significant in many African settings – due in part to urban women's time poverty as they juggle paid labour and unpaid reproductive roles, limiting time for food preparation (Tacoli 2017). Maintaining extra-household social relationships and kinship networks that provide food through remittances and exchange is another important means of food access (Mackay 2019). Also helping to mitigate poverty and food insecurity are women's savings clubs. Varying in nature and form, these are an important means of pooling resources and spreading risk in conditions of economic precarity (James 2015). When market and non-market access strategies fail to provide sufficient food, common intra-household strategies are to reduce food consumption or limit dietary diversity. Studies in a number of urban settings in African and other low-income contexts have shown how adopting such strategies 'disproportionately affects women, as they are often the last ones to eat and tend to forego food to ensure children have enough' (Tacoli 2017, p. 4).

The fact that employment and income sources are key to household food access and food security in urban settings makes it important to understand emerging changes to female labour market participation in African cities. As Alice Evans (2014) has argued in her study of Kitwe, economic insecurity has led to increased social acceptance of women engaging in paid work in stereotypically male occupations, and to changing aspirations for women and girls in terms of education and career. Family forms are also changing. The increasing proportion of households that are female-headed is driven by 'changes in marriage behavior, family formation, health, and education' (Milazzo and van de Walle 2015, p. 1). Milazzo and van de Walle's quantitative analysis of survey data for 24 African countries shows that although there is an association of female headship with poverty, poverty has been falling faster in female-headed than male-headed households, albeit with considerable variation across different categories and definitions of female headship (e.g. widowed, divorced, abandoned, or single women; *de jure* and *de facto* female heads). Heterogeneity within the category of female-headed households does not negate the role of gender in shaping poverty and food security; rather, it highlights the significance of social reproduction, including how family form and gendered labour are changing in response to broader political-economic and social change.

Feminist scholars have long called for a centring of social reproduction in thinking about cities (Mackenzie 1989; Jarvis, Cloke, and Kantor 2009; Peake and Rieker 2013; Chant and McIlwaine 2016). Cindi

Katz described social reproduction as 'the fleshy, messy, and indeterminate stuff of everyday life' (2001, p. 711). Social reproduction, in that broad definition, encompasses the 'acquisition and distribution of the means of existence, including food, shelter, clothing and health care' (p. 711); the reproduction of a skilled and differentiated labour force; and reproduction of society as a whole, including social forms, relations, and practices. Reproductive labour, then, encompasses the unpaid labour of feeding and caring for families in the domestic sphere, and paid labour in the care and health sectors, such as domestic work or child and elder care. The distinction between productive and reproductive labour is, in reality, often blurred. Indeed, as Jamie Winders and Barbara Ellen Smith (2019) suggest, if research in early feminist geography had been more focused on the global South, 'the very notion of production and social reproduction as distinct spheres of daily life would have made less sense as an ontological starting point for theorizing their relationship' (p. 883).

As activities and as spaces, production and reproduction certainly blur in African urban contexts. Day-to-day activity for many people in these urban environments is organized around basic food provisioning for their families and the imperative of avoiding hunger. Women's opportunities for remunerated occupations are constrained by social expectations of performing unpaid domestic and care work, including purchasing and preparing food, and yet they do engage in income-earning activities – paid employment, formal and informal trading activity, and various forms of home-based production – many of these involving food. They also work to maintain informal institutions of non-market exchange and social support. Their productive and reproductive labour supports not only families but communities, in contexts where other forms of social protection are weak or non-existent. We join other feminist scholars who argue that feminist interventions need 'to engage more fully with the everyday struggles of living and working that animate urban dwellers and the complex scaffolding upon which the vast majority of the world's women living in cities, the working poor, pin their hopes and dreams' (Peake 2016, p. 225).

The Consuming Urban Poverty (CUP) Project: Research Methods and Researcher Positionality

Led by the African Centre for Cities (ACC) at the University of Cape Town, and funded by the UK's Economic and Social Research Council (ESRC) and Department for International Development (DFID), CUP was a coordinated programme of collaborative South–South research conducted jointly with the Copperbelt University in Zambia, the University of Zimbabwe, and the Kisumu Local Interaction Platform

(KLIP) in Kenya, each of which had already established links with ACC.[1] The specific research activities were developed together with the partner organization in each city within an overarching framework and set of objectives (Battersby and Watson 2018). The guiding motivation was that 'important contributions to debates on urbanization in sub-Saharan Africa, the nature of urban poverty, and the relationship between governance, poverty and the spatial characteristics of cities and towns in the region can be made through a focus on urban food systems and the dynamics of urban food poverty' (CUP website 2018).

In designing and carrying out their programme of research, CUP adopted the broad definition of food security articulated by FAO (2006). This sets out the four pillars of food security as availability, access, utilization, and stability. Food security requires not only the physical availability of food but the ability of households and individuals to access food, by financial or other forms of entitlement. It further depends on people's ability to secure overall nutritional and physiological well-being, including non-food inputs such as clean water and sanitation. Finally, food availability and access should be secure over time, including the ability to withstand seasonal fluctuations and sudden environmental or price shocks (Battersby and Haysom 2018). This conceptualization of food security extends its scope beyond the household to consideration of wider city, national, and international scales: in other words, to food systems. There were four main methodological components to the larger CUP project (Battersby and Watson 2018): retail mapping and surveying of food outlets; a value chain analysis for particular food commodities; a combination of in-depth household interviews and a questionnaire survey to gather information on food access and food poverty; and an assessment of food system governance in each of the three cities.

For the household-level components of the project, each city's team adopted a slightly different approach when selecting neighbourhoods to sample. The Kisumu sample, to capture a range of socio-economic status and settlement types, covered a larger number of distinct neighbourhoods (informal/formal and central/peri-urban). In Kitwe, two neighbourhoods were selected, specifically on the basis of being poorer parts of the city. The Epworth team studied three wards – one poor, one better off, and one in between – although deep and widespread poverty makes the differences amongst them relatively small. None of the samples, for either the quantitative or qualitative components, should be taken as representative of their city as a whole. They are, however, illustrative of conditions in the communities from which they were drawn, and they depict residents' engagement with the wider urban food system.

The overall CUP project did not adopt an explicitly feminist approach, although gender-based variables were included in the survey instrument and gender-related questions were included in the qualitative interviews. Neither of the present authors was directly involved in the CUP project's conceptualization, design, or collection of primary data. One of us (Belinda Dodson) was invited in the later stages of the project cycle to conduct a deeper gender analysis building on preliminary CUP findings, which had indicated the importance of gender in urban food systems and food poverty. In that role, Belinda was granted access to the full suite of CUP data as well as to analysis and reporting by other members of the CUP team. Belinda invited Liam Riley to co-author this chapter, based on long-term collaboration in other work on gender and urban food security in Southern Africa. This places us in a rather unusual researcher position, as post hoc additions to an established research team. We see our contribution as enriching and extending the conclusions drawn from overall and city-specific CUP findings (as presented in Battersby and Watson 2018; Opiyo et al. 2018; Tawodzera and Chigumira 2019; Tawodzera and Chileshe 2019). Our analysis highlights the gendered nature of both the urban food system and lived experience of urban food poverty. We base our analysis primarily on the qualitative household interviews carried out by CUP's local research assistants. We acknowledge the privilege of being granted access to the interview notes and transcripts and we give full credit to the local partners, research assistants, and interview participants for the analytical value and empirical richness of the material.

CUP teams collected survey and interview data in 2016 and 2017. Surveys collected data from 879 households in Kitwe, 841 in Kisumu, and 483 in Epworth, employing standardized household food security measures and using randomized spatial sampling within the selected neighbourhoods (starting at a software-generated random point and then sampling every nth house, based on the size of the community). Prior to the surveys, CUP researchers carried out in-depth interviews, in the same selected neighbourhoods. Local CUP project teams conducted interviews with people from 50 households in Kisumu, 15 households in Epworth, and 15 households in Kitwe. In keeping with the general guidelines of the project, the interview question guide was designed and adapted for each setting. Interview samples were based on a mix of spatial and snowball strategies.

Initially, household interview findings were used by the larger CUP teams to guide survey design and analysis as well as to provide information on the socio-cultural context in which food systems and food security are situated. In conducting our gender analysis, we have used the primary data from these interviews to examine the role of gender and gender relations in everyday experiences of food access and food poverty. In Kisumu and Kitwe, the household interviews conducted by the local project team were all with women. In Epworth, although most interviewees were women, a few men were interviewed too. The selection of women as respondents reflects local norms in which women are associated with knowledge about their households' food access and consumption practices, but it does risk reinforcing assumptions and stereotypes about gender roles and may have meant that certain experiences and perspectives were missed. Collection of survey and interview data recorded people's gender in binary terms as male or female (or as 'refused to answer' or 'missing') and asked about individuals' relation to the 'head of household', thus allowing some exploration of gender relations but within a particular normative framework. A similar caveat applies to the CUP project definition of a household as people sharing a dwelling for at least six months in a year. That definition acknowledges some complexity but might under-recognize the fluidity of household relationships, form, and temporality. CUP interview findings nevertheless reveal some of the complexity and fluidity of family and gender relations, and the importance of those relations in food security and food systems in these three African secondary cities.

Urban Food Systems and Food Insecurity in Kitwe, Kisumu, and Epworth

The three cities selected as locations for CUP research possess a number of historical and contemporary similarities. Each bears the spatial and institutional imprint of British colonial legacies of urban planning and administration. Today, in addition to high levels of urban poverty common to many African cities, they also share problems of weak urban governance. Furthermore, '[d]evelopment policy and governance dynamics in all three are characterized by a degree of anti-urbanism and a reluctance to accept or deal with problems of urbanization and urban poverty, leaving poor urban residents vulnerable to economic and environmental shocks, including state-led evictions' (Battersby and Watson 2018, p. 10). Within these broad similarities, the cities differ in their respective economic bases and development trajectories (Battersby and Watson 2018).

- Located in western Kenya on the shores of Lake Victoria, Kisumu is Kenya's third-largest city after Nairobi and Mombasa. It is an important trade and services centre. It has high levels of poverty and unemployment and over 50% of the working population are employed in the informal sector. Roughly 60% of the population live in informal settlements, but the city also has new private estates occupied by middle- and high-income residents. The food system is centred on several food markets, both formal and informal, and there is a growing number of supermarkets. Street traders and mobile vendors are also significant actors, despite official restrictions on their activity.

- Kitwe is a mining town in the Copperbelt region of Zambia. Privatization of the mines in the 1990s saw mass layoffs and Kitwe's economy remains vulnerable to fluctuating copper prices. An estimated 50% of the population live in informal settlements. The food system includes one main market, Chisokone Market, along with smaller formal and informal markets, street traders, and mobile vendors. There are also several supermarkets, for example the South African chain Shoprite. City authorities have tried to relocate traders from the overcrowded central market to new markets in more peripheral sites.
- Epworth expanded to become a satellite town to Zimbabwe's capital, Harare, following the lifting of racially exclusionary policies of urban influx control after the country's independence in 1980. About 70% of Epworth's population live in informal conditions and only 23% of the working population were in formal employment in 2014. Although there are formal shopping precincts, informal trade including street vendors and 'tuck-shops' dominates retail food trade. There are numerous supermarkets and other formal shops in Harare, approximately 20 kilometres away.

Quantitative and qualitative CUP findings point to differences in how food systems are engaged and experienced by men and women, and by people in different types of households. One key finding was a high prevalence of food insecurity, found to be worse for female-headed households on a number of food access, dietary diversity, and lived poverty indicators. Table 10.1 shows household-level CUP data comparing female-centred with nuclear households in relation to Household Food Insecurity Access Prevalence (HFIAP).[2] HFIAP categorizes households into four levels of food insecurity: food secure, and mildly, moderately, or severely food insecure. Households are determined as increasingly food insecure as they experience conditions of food insecurity more severely and more frequently.

In addition to higher overall prevalence of food insecurity in the Epworth and Kitwe samples, severe food insecurity is higher in female-centred households in all three cases. The Kisumu sample's better aggregate food security status partly reflects the inclusion of well-off neighbourhoods. But, as detailed below, gender intersects and interacts with other demographic and social factors to shape the extent and experience of food insecurity. Marital status, household composition, age, and stage in the lifecourse all affect people's food access and entitlements, including intrahousehold bargaining over money, food, and other resources.

Table 10.1 HFIAP Food security status by household type

CUP Cities	Household Type	% Severely Food Insecure	% Moderately Food Insecure	% Mildly Food Insecure	% Food Secure
Kitwe	Female-Centred	86.4	9.4	1.0	3.1
	Nuclear	75.0	13.7	3.4	7.9
Kisumu	Female-Centred	57.0	20.7	6.5	15.8
	Nuclear	45.0	25.9	9.0	20.1
Epworth	Female-Centred	74.7	16.8	5.3	3.2
	Nuclear	65.7	21.2	4.2	8.8

A second key finding of the overall CUP project was the importance of food purchase as the main source of food, using a range of formal and informal retail outlets. In light of the importance of purchasing as the predominant means of accessing food in these three cities, income is crucial to urban households' food security, so it is important to understand people's employment and occupations. Informal food retail was found to be essential to food purchasing by poorer households, as well as providing a source of employment and income – particularly for women (Skinner 2018).

A third conclusion of CUP research was that urban plans, policies, and laws governing urban food systems and spaces can negatively affect food access and food security (Smit 2018). Interventions restricting the informal sector in particular have gender-uneven impact. The qualitative analysis presented in the remainder of this chapter reveals gendered experiences of food security and food systems, rooted in gender roles, family and intra-household dynamics, and social reproduction.

Lived Urban Geographies of Food Access and Food Poverty in Kitwe, Kisumu, and Epworth

Gender-informed interpretation of CUP findings shows the importance of avoiding over-generalization on the basis of an aggregate comparison between nuclear and female-headed households, even though that can form a starting point. Female-headed households are heterogeneous, incorporating younger and older women, women with or without children, households with or without other adult members, and various forms of multi-generational households. Across all household types, women's knowledge, labour, and social networks are essential to everyday household reproduction through food, and also to strategies for household survival in conditions of extreme poverty. Describing these experiences and examining how people navigate the food system reveals the gendered social forms and practices that reproduce life, family, and labour in conditions of urban precarity, along with the ways urban food systems can act to reproduce gendered inequality.

Marital Status, Household Form, and Gendered Occupations

In all three places, interviewees included several widows, divorced women, and other female household heads, as well as women with male partners in nuclear and extended households. Interviews did not explore whether poverty and food insecurity had led to marital breakdown, but, in several cases, marital breakdown or death of a spouse had certainly exacerbated food poverty. In other cases, households where there was no male partner were better off, a finding consistent with research showing the growing prevalence of female-headed households in Africa without an inevitable association with increasing poverty (Milazzo and van de Walle 2015).

Widows described particularly precarious circumstances. One widow in Kitwe earned income from selling charcoal. Two other members of her household contributed to household income: a daughter who worked in a shop and a son who worked as a security guard. A further two adults and a grandchild made up her household of six. Despite the household having multiple income sources, she 'does not budget because food is bought on a daily basis depending on income.' Still worse off was another widow in the same neighbourhood. In addition to herself, her household consisted of two adult children and two grandchildren. Her son was in prison and her daughter physically disabled, so neither was able to contribute to household income. This woman earned a livelihood by trading in agricultural produce purchased at a market almost 100 kilometres away. She described monthly shortfalls in income to meet even basic food needs, noting that the household routinely did not eat a morning meal. A widow in Kisumu observed that she was the head of her household of five, which included an adult son and daughter and two grandchildren. She described herself as the sole provider and complained about the high cost of food. This meant that she 'eats only once a day; twice is a blessing.'

Not all female-headed households were food insecure or living in conditions of economic precarity. Amongst the best-off households across the three city settings were a small number of multi-generational, female-centred households in Kisumu comprising older women and their adult daughters, some also with other relatives, and often including one or more children. These households were located in more affluent Kisumu neighbourhoods (a type of neighbourhood not included in the Kitwe or Epworth samples). Interviewees identified the older women as heads of these households, but their adult daughters as the main income earners. These younger women, unlike most women in the Kisumu sample, had career-building jobs such as accountant or pharmacist, with higher and more reliable income.

Being in a nuclear household was no guarantee of food security. Married women indicated varying economic circumstances and experiences. For some, being occupied with childcare and other domestic responsibilities constrained their own income-earning opportunities, making them entirely reliant on a spouse's income. One Kitwe woman's husband sold flowers and plants at a local shopping centre. This work provided a low and unreliable income yet was the sole source of money for buying food and other necessities for the couple and their four children. Another mother from Kitwe, aged 18 and with a baby, depended on her husband's erratic earnings from various forms of casual labour. She herself had given up informal doorstep food-vending, saying that she did not get enough money from her husband to buy goods like charcoal or *kapenta* (a sardine-like fish) for resale.

Households with higher food security also had more reliable income from one or more sources. Many of these were nuclear or extended-family households with a male 'breadwinner'. One Kitwe woman's husband was a street vendor in Lusaka, the distant capital city of Zambia; another's husband rented out five 'cabins' on their property. These households, too, experienced occasional food shortages, but were more able to budget, buy food in bigger quantities, and eat three meals a day – which in this context counts as being relatively food secure. Some of these better-off households were able to refrigerate food, allowing healthier diets as well as the possibility for economizing through buying food in greater bulk instead of daily purchases of perishables.

Women in nuclear households commonly reported gendered domestic roles of a male household head being responsible for providing income to purchase food, but actual food purchasing, storage, and preparation being the domain of the wife/mother, often expressed in terms of 'my duty as a woman'. This was true even when women earned income independently through their own economic activity. Women in nuclear households with more than one income earner still reported that rising food prices and insufficient or inconsistent income were constraints to household food security. As reported by a married woman with children in Kisumu: 'Food is highly available; the only challenge is the finances but getting goods in the market is not a problem.'

One Epworth couple earned income from the male adult's self-employment as a panel beater and his female partner's casual employment as a domestic worker. This allowed them to purchase a sufficient quantity and considerable variety of foods, including staples like maize meal and cooking oil as well as occasional 'luxuries' like meat, supplementing this with vegetables grown in their garden. Despite their current relative economic security, this female respondent shared recollections of previous periods of considerable food insecurity. Things had been particularly tough in 2012, when their child was born and both the woman and her husband had been temporarily unemployed. They had also experienced more recent, short-term periods of food insecurity when the husband's business had slowed down, but they had been able to weather these by drawing on family and social networks for support.

Perhaps the most destitute household, counter to the aggregate pattern of nuclear households being better off, was that of a married couple in

Epworth with six children. The husband was experiencing difficulty finding work in his usual occupation as a driver. The woman had worked as a trader, but all her goods had been stolen and she had no capital to restart her business. The household relied almost entirely on food donations from neighbours and a local church, along with any income the woman earned from casual work like cleaning houses or doing laundry.

Self-employment and casual employment dominated the employment profile in all three settings. Reported occupations ranged from a small number in professions such as being a pharmacist (female), accountant (female), teacher (female), or army employee (male); through occupations like house cleaning, laundry, child care, or hairdressing (women) and gardening, security, construction, driving, or panel beating (men); to various forms and scales of informal sector trading, typically in food but sometimes in other goods (women and men). Although there has been some change, occupations still reflect gendered labour markets along with social norms about what are regarded as appropriate economic activities for men and women. A number of women gave their occupation as 'housewife' – typically women with pre-school children. In addition to performing childcare, cooking, and other reproductive labour, women describing themselves as housewives were sometimes producing food for domestic consumption, for example by growing vegetables, sometimes selling a small surplus. This could be in a home garden or on nearby land, through a variety of formal and informal arrangements by which people accessed or simply appropriated urban or peri-urban land for cultivation.

Another clear finding is extensive involvement of women in the urban food economy as sellers as well as buyers. In all three cities, food vending was an important source of income for women. Women reported selling food they had produced themselves, such as poultry, maize, or vegetables, or food that they had purchased in bulk and then resold in smaller quantities, such as chicken pieces, fish, eggs, or tomatoes. There is a gendered geography to food vending. Some women, especially those with young children, sold food and related goods from rudimentary stalls on their own doorsteps; others sold their wares from stalls in informal and more formal local markets. One female household head, a widow in Kitwe, travelled to a distant border town to take advantage of price differentials in buying and selling agricultural produce, but women were more likely to engage in local trading activity. Despite the importance of food vending as occupation and livelihood, traders in all three settings reported actions by city officials that sometimes made their vending activities difficult or even impossible, for example through regulatory constraints or prohibitions on street vending.

In all three places, macroeconomic conditions and insecure employment mean that many households struggle to secure the most basic means of subsistence. As outlined in the section below, this affects how individuals and households engage in the urban food system, and what measures they adopt when they are unable to access sufficient food through market-based means.

Food Procurement and Access

Households practise an array of livelihood and food procurement strategies. In all three places, buying and preparing food was primarily described as women's responsibility, expressed as part of their roles as wives and mothers. Men were typically described in 'breadwinner' terms, but, with very few exceptions, it was reported that purchasing or otherwise procuring food was the responsibility of women. As one married woman in Kisumu described: 'My husband gives the money; I decide what to do with the money as long as we do not sleep hungry.' In female-centred households, it was often just one adult female responsible for both roles, unless there were additional economically active adult members. Across all three locations, women's experience and expertise in food buying was portrayed as an asset, even a source of pride. 'Because of my experience, I know where to get better bargains,' said another Kisumu woman.

In addition to the central role of women, five further features stand out in respondents' food procurement practices and engagement in these urban food systems. First is the predominance of food purchasing. Even in cases where households practised some food production, most of the food consumed was bought. A second key finding, related to the precarity of income and livelihoods, was that many households purchase small quantities of food on a day-to-day basis, for example the small packages of maize meal known as *pamelas* in Zambia. The savings that could be made by buying in bulk from supermarkets or wholesalers were simply out of financial reach for most households, making local, informal outlets especially important. A third pattern evident across all three sites was the wide variety of sources of food purchases, even within the food purchasing practices of individual households. Sources included: street vendors and roadside stalls; neighbourhood kiosks; informal markets; formal markets such as those managed by municipal authorities; and formal grocery stores, supermarkets, and wholesalers. There were synergies amongst these different outlets, for example informal sellers sourcing stock from supermarket and other formal sources. A fourth notable characteristic is the importance of the immediate local community in people's food purchasing, with convenience, affordability, and credit, along with transport costs and time poverty, expressed as influencing factors. A fifth feature was supplementation of food purchasing by other, non-market sources of food, although these sources were less frequent or common than might have been expected. Amongst the sources described were: own production, either in urban or in rural areas; food remittances from relatives; gathering wild foods of various types; borrowing food from neighbours or relatives; and receiving food donations from organizations such as churches or NGOs.

Households that had the time and money to do so diversified their food sources as a means of stretching food budgets: travelling to more distant, cheaper markets if they could pay for transport; or buying food in larger quantities from markets, supermarkets, or wholesalers. The Kitwe widow who travelled to a distant border town for her trading business also purchased food there for her own household's consumption: 'I purchase the bulk of dry foods from the border when I go for business ... I purchase mealie meal, salt and sugar and then for protein I get caterpillars (*ifishimu*).' Some households did not have enough money for transport, as expressed by a married woman in Epworth: 'Due to limited amount of money, we cannot afford to board a bus and go to buy at OK supermarket [in Harare].' Mobility, and its affordability, thus becomes a crucial element in household food strategizing.

There was a temporality as well as a spatiality to these purchasing patterns: buying from supermarkets or formal markets at times when households had enough money, but making smaller purchases from local informal vendors, often on credit, when money was short. As one Kitwe woman noted:

> During lean times I am able to borrow a *pamela* [small package of maize meal] and a charcoal small pack on a daily basis from a *tuntemba* [informal food stall] and this is paid for after my husband comes home from work.

Personal relationships of familiarity and trust were identified as important in getting bargains or being granted credit, typically with informal food vendors operating in the local vicinity.

In all three locations, there were only a few households producing their own food, either in urban or rural areas. Some women reported growing vegetables in kitchen gardens, providing food for family consumption and sometimes a source of income:

> I have a garden in which I grow vegetables and tomatoes. That's what I eat my *sadza* [maize meal] with in the afternoon. ... I cannot say I am employed. I just have my garden from which I sell produce to my neighbours when they ask me to. (Married woman, Epworth)

A number of interviewees expressed regret at not having access to land on which to grow food. Although lack of land was the most commonly cited reason, other factors cited for not practising urban agriculture included time constraints or fear that people would steal what was grown. As reported in the wider CUP findings, urban expansion and land pressure, insecure land tenure, and contradictory and inconsistently applied city policies on urban and peri-urban agriculture were undermining households' ability to engage in food production (Battersby and Watson 2018).

A few households received food from relatives in rural areas, providing an important supplementary food source. As related by another married woman in Epworth:

> I have an aunty in Gokwe who sometimes sends us maize, sugar cane and pumpkins. This helps a lot in food security of the household. For example, once they send even a bucket of maize then you know you are covered you don't have to buy mealie meal. The money saved from not buying that particular item can allow you to buy food items such as fruit which you were no longer eating as a result of having a tight budget.

Some respondents reported that this had been a more common food source for their household in the past. Their explanations combined family factors with wider political, economic, and environmental causes. They attributed the decline to drought affecting harvests, or to illness, ageing, or death of the rural-based relatives who had previously sent food. In Kisumu, one woman explained that they no longer received food from the village because farms had been destroyed in political clashes. A retired man in Epworth said simply: 'As of these current years, things are very tight such that you cannot expect anything from there.'

Several respondents in each city reported borrowing food from relatives or neighbours in town, acknowledging that these loans were sometimes tacitly understood to be donations. Others reported lending or giving food to neighbours in need, either directly or by inviting neighbours or neighbours' children to share meals. In Kitwe and Epworth, interviewees said they were now sharing food with neighbours less frequently, as they did not have surplus food to spare. An Epworth woman from an extremely food insecure household reported that she had become 'scared of borrowing because I know I have no way to ensure repayment [and] I don't want bad blood with my neighbours.'

People reported a number of additional social and behavioural strategies for dealing with chronic or periodic food poverty and hunger. Amongst the adaptations in food purchase and consumption at the household scale were: cutting out all but the cheapest foods; eating only one or two meals a day; drinking sugary tea as a meal substitute; and preparing food in particular ways to give a feeling of fullness. A married woman from a nuclear family in Kitwe, comprising herself, her husband and four children, described the measures that she and her husband adopted:

> Many times breakfast is skipped so that I am able to concentrate on lunch. The supper is brought in by my husband after his sales. The usual meal is *nshima* [maize meal] and vegetables but when there is money I am able to buy *kapenta* [dried fish]. When there is no money I give my children sugar solution called *zigolo* or plain tea without any accompaniment and this keeps them calm until I am able to cook some lunch.

A widow in Kisumu, head of a three-generation household but the sole income earner, said: 'Today I have had tea and I may not eat again.' The most abject household of all was one in Epworth, a married couple with six children:

> Like right now, my husband is not going to work so in the morning we might not have anything to eat and strive to get to midday on empty stomachs then we roast peas then we drink water afterwards and wait for *sadza* in the evening if we have mealie meal. In the case that we have no mealie meal, we eat roasted peas again for supper then we go to sleep.

She cooked the peas with a lot of salt to stimulate thirst, because drinking water creates a feeling of fullness. The family had not been able to purchase food for the previous six months, and the peas had been donated by a neighbour 'after they had noticed that we were almost starving.' They also received a monthly food hamper from a church. In Epworth and Kisumu, a few interviewees reported receiving such donations from faith-based or other NGOs.

Parents reported making sacrifices when money and food were scarce. As a married woman in Kisumu expressed it: 'At times we are forced to skip meals but we have to ensure the children are provided for.' Taking in the children of relatives or neighbours, or sending their own children to stay or eat with relatives or neighbours, were other reported strategies. A male respondent in Epworth, whose multi-generational household of seven consisted of himself, his wife, their three children, his elderly father, and a nephew, relieved the pressure on his household by accessing food away from home:

> Moreover, I don't eat [at home] most of the times because I want my family to have enough food to eat. Sometimes I subtract myself from certain meals so that they have something to eat... sometimes I can go out with my friends and have my meal [with them] in the evening.

A social strategy practised by women in all three settings was that of informal savings groups. Members contribute a certain amount each week or month and the pooled amount either goes to one member to tide them over in an emergency, or to buy food in bulk to be divided amongst the group. These were described as an important means of mitigating household poverty and food insecurity. Some women, like this one in Epworth, reported that they were now 'too poor to join a savings group':

> We used to do *mikando* [savings group] in groups of 10 for money to buy groceries. We used to pay $20 each per month and we used to share on the 24th of December. The food stuffs used to be bought at Reuben shops here in Epworth. There is a wholesale there. We have since stopped the club last year because people no longer had enough money to spare and this has negatively impacted our food access.

The procurement strategies and mitigation measures described above reveal gendered roles, relations, and practices within households, along with gendered social networks outside the household, that sustain urban families and communities. To feed their families, women navigate food retail systems, intra-household relations, and social and kinship networks. Even market-based strategies, such as buying food on credit from local vendors, are embedded in social relations of trust and familiarity. The work required to maintain those social networks and relationships, or to devise alternative strategies when networks and relationships fail or weaken, constitutes an important but overlooked part of urban food systems.

Conclusion

Looking at cities through food highlights the gendered labour, social relations, mobilities, and spaces essential to urban social reproduction. Understanding urban food systems and food security requires understanding how households engage these systems, and how urban economic, political, and social contexts affect that engagement. It also requires examination of intra-household dynamics and relations that shape how households strategically allocate labour, time, financial resources, and social capital to secure food for their members. Increased urbanization of African societies and economies is bringing changes to household forms, to gendered patterns of employment, and to social relations of reciprocity and exchange. These interact in multiple ways to affect urban food security at household, community, and city scales.

CUP findings indicate changes in urban family structures and labour markets and how these are affecting gendered experiences of food insecurity. Quantitative food security indicators showed female-headed households to be significantly less food secure in all three study locations. Information collected in the interviews highlighted diversity within the category of female-headed households. Widows in all three settings reported extreme food insecurity and precarious livelihoods. By contrast, there were a few multi-generational female-headed households amongst the Kisumu interviews, with women in skilled occupations, who were food secure. This suggests changes in female incorporation into urban labour markets, possibly reflecting wider changes to gender norms and stereotypes. Sylvia Chant (2015) asked the provocative question of whether some forms of 'female-headedness might be considered as an asset, with a role to play in making cities of the future more gender-equitable' (p. 21). Tentatively, based on only a few female-headed households in Kisumu, CUP findings support Chant's hypothesis, signalling that the widely observed increase in female headship in African urban settings may have some positive outcomes.

Urban food systems themselves act to reproduce socio-economic inequality. Physical availability and abundant choice of foodstuffs in diverse retail outlets is an emerging characteristic of African urban food systems. CUP retail and household surveys identified the presence and utilization of a diversity of food retail outlets in these secondary cities, across a spectrum from micro-scale informal food stalls to international supermarket chains (Battersby and Watson 2018). Urban residents' ability to access those sources, however, was highly uneven. Intra-urban mobility to draw on multiple retail outlets was one strategy employed to access more diverse and cheaper food, but the transportation required was financially inaccessible to the poorest households. Urban infrastructure, the spatial location of housing and food markets, and affordable public transportation can thus be seen as essential to urban food system functioning to support household and city-scale food security. Women, who remain primarily responsible for the reproductive labour of family food provisioning and preparation, stand to benefit from such investments.

Also evident in CUP findings, and common across African cities, was women's significant activity as vendors in informal food retail. Informal food vending provided an essential if precarious source of livelihood for many women. Some women, notably in Kitwe and Epworth, had involuntarily ceased their food vending activity because they no longer had enough money to buy goods for sale or because of official restrictions on certain types of informal food trade. In each of the three CUP cities, elite official imaginaries perceived certain forms of informal food vending as detracting from the idealized modern city that urban authorities seek to reproduce (Battersby and Watson 2018). Urban food system governance that fails to support diverse informal food retailing effectively discriminates against women and the urban poor, and risks damaging the very means by which households and communities are sustained.

Although urban food systems are dominated by money-based modes of food procurement, there are also non-market sources of food, sustained by social networks and kinship ties. CUP survey findings showed non-market food sources to be limited in extent, but interview findings revealed that they are crucial to some households. Through borrowing and lending food amongst neighbours, membership in savings clubs, and maintaining relationships with relatives in rural and urban areas, women supplement the food security of their own and other households. This demonstrates the embeddedness of urban food security in social networks and relationships, whose maintenance constitutes a form of reproductive labour. Social relations underpinning food security extend beyond city boundaries, for example in bi-directional mobilities of people and food through kinship-based rural-urban linkages. Some interviewees reported these networks and linkages to be in decline, reflecting not only macroeconomic conditions but also weakening rural ties over time, threatening the sustainability of this important social safety net.

CUP findings for Kitwe, Kisumu, and Epworth show strong similarity with other research on gender and urban food security in African contexts, for example the African Food Security Urban Network (AFSUN) (Dodson, Chiweza, and Riley 2012) and recent work by Stevano (2019) in Mozambique and Mackay (2019) in Uganda. They show how gender norms can be challenged in cities (Evans 2014), providing greater social acceptance of women engaging in paid labour, but also how women's family reproductive responsibilities persist, adding to time burdens and restricting women's economic participation. They emphasize the importance of women as actors in urban food systems, both as vendors and as purchasers who 'buy wisely' to meet their families' food needs. They highlight the significance of social relations within and beyond the household and the role of food in social reproduction. Applying a gender lens to urban food security exposes contradictions, gaps, and inadequacies in urban food systems and the gender blindness of urban food governance. Understanding urban food security requires attention to people's living and working conditions in particular places, including how these are changing over time in conditions of rapid urbanization, economic uncertainty, and weak governance. Without such understanding, 'the ways in which gender roles, responsibilities, and expectations are normalized and often the root of inequality in terms of food access and security' (Riley and Hovorka 2015, p. 337) are likely to persist, reproducing gender inequality, household food poverty, and unequal inter- and intra-urban geographies of food insecurity.

Acknowledgements

Belinda Dodson is grateful to Jane Battersby and Vanessa Watson for inviting her to do a gender analysis of the Consuming Urban Poverty project data, including participation in a CUP workshop in Cape Town in July 2018. We owe an enormous debt of gratitude to the entire CUP research team, including partners in the three study cities and in Cape Town, especially the research assistants who collected interview data in Kitwe, Kisumu, and Epworth (named in Note 1). The project was funded by the Economic and Social Research Council (UK) and the UK Department for International Development under grant number ES/L008610/1, with the full title 'Governing Food Systems to Alleviate Poverty in Secondary Cities in Africa', within the ESRC-DFID Joint Fund for Poverty Alleviation Research (Poverty in Urban Spaces theme).

Notes

1   A number of people were involved in the design and execution of the CUP project. At ACC in Cape Town, the Principal Investigator was Vanessa Watson. Jane Battersby, Gareth Haysom, Susan Parnell, and Warren Smit were other key members. Project team leaders in the partner cities were Francis Muwowo, Owen Sichone, and Niraj Jain in Kitwe, Zambia; Paul Opiyo, George Wagah, Patrick Hayombe, and Stephen Gaya Agong in Kisumu, Kenya; and Easther Chigumira and Godfrey Tawodzera in Epworth, Zimbabwe. Although there were more men than women in leadership roles, the research assistants, especially those who carried out the interviews, were a more gender-balanced group. Eva Mazala did all the household interviews in Kitwe, working with Owen Sichone. In Kisumu, Paul Opiyo did the interviews together with research assistants Edith Akunja and David Owuor. The Epworth interviews were carried out by a team of five: Heather Chachona, Amanda Dendera, Sithabile Mbambo, Gift Mudozori, and Anesu Nyamba.

2   Female-centred households were defined as those in which there was a female 'head', along with any children and possibly other adult members, but no male partner permanently present. Nuclear refers to households with a male and female adult member, along with any children.

References

Battersby, J. and Haysom, G. (2018). Linking urban food security, urban food systems, poverty, and urbanization. In: *Urban Food Systems Governance and Poverty in African Cities* (eds. J. Battersby and V. Watson), 56–67. New York: Routledge.

Battersby, J. and Watson, V. eds. (2018). *Urban Food Systems Governance and Poverty in African Cities*. New York: Routledge.

Chant, S. (2014). Exploring the 'feminisation of poverty' in relation to women's work and home-based enterprise in slums of the global south. *International Journal of Gender and Entrepreneurship* 6 (3): 296–316.

Chant, S. (2015). Female household headship as an asset? Interrogating the intersections of urbanisation, gender, and domestic transformations. In: *Gender, Asset Accumulation and Just Cities: Pathways to Transformation* (ed. C.O.N. Moser), 21–39. London: Routledge.

Chant, S. and McIlwaine, C. (2016). *Cities, Slums and Gender in the Global South: Towards a Feminised Urban Future*. New York: Routledge.

Cock, J. (2016). A feminist response to the food crisis in contemporary South Africa. *Agenda* 30 (1): 121–132.

Consuming Urban Poverty (CUP) (2018). *Consuming Urban Poverty: Food systems planning and governance in Africa's secondary cities*. https://consumingurbanpoverty.wordpress.com (accessed 18 December 2018)

Consuming Urban Poverty (2019). *Incorporating Food into Urban Planning: A Toolkit for Planning Educators in Africa*. Cape Town: African Centre for Cities.

Crush, J. and Frayne, B. (2011). Supermarket expansion and the informal food economy in Southern African cities: Implications for urban food security. *Journal of Southern African Studies* 37 (4): 781–807.

Crush, J., Frayne, B., and Pendleton, W. (2012). The crisis of food insecurity in African cities. *Journal of Hunger & Environmental Nutrition* 7 (2–3): 271–292.

Dodson, B., Chiweza, A., and Riley, L. (2012). *Gender and Food Insecurity in Southern African Cities*. Urban Food Security Series No. 10, Kingston, ON and Cape Town: African Food Security Urban Network.

Evans, A. (2014). 'Women can do what men can do': The causes and consequences of growing flexibility in gender divisions of labour in Kitwe, Zambia. *Journal of Southern African Studies* 40 (5): 981–998.

Floro, M. and Swain, R. (2013). Food security, gender, and occupational choice among urban low-income households. *World Development* 42: 89–99.

Food and Agriculture Organization (FAO) (2006). Food security. *Policy Brief June 2006, Issue 2*. www.fao.org/forestry/13128-0e6f36f27e-0091055bec28ebe830f46b3.pdf.

Hovorka, A., de Zeeuw, H., and Njenga, M. eds. (2009). *Women Feeding Cities: Mainstreaming Gender in Urban Agriculture and Food Security*. Rugby, UK: Practical Action Publishing.

International Labour Organization (ILO) and WIEGO (2013). *Women and Men in the Informal Economy: A Statistical Picture*. Second Edition. Geneva: International Labour Organization.

James, D. (2015). 'Women use their strength in the house': Savings clubs in an Mpumalanga village. *Journal of Southern African Studies* 41 (5): 1035–1052.

Jarvis, H., Cloke, J. and Kantor, P. eds. (2009). *Cities and Gender*. London: Routledge.

Katz, C. (2001). Vagabond capitalism and the necessity of social reproduction. *Antipode* 33 (4): 709–728.

King, A. (2016). Access to opportunity: A case study of street food vendors in Ghana's urban informal economy. In: *Gender and Food: From Production to Consumption and After* (eds. M. Texler Segal and V. Demos), 65–86. Bingley, UK: Emerald Group Publishing.

Levin, C., Ruel, M., Morris, S. et al. (1999). Working women in an urban setting: Traders, vendors and food security in Accra. *World Development* 27 (11): 1977–1991.

Mackay, H. (2019). Food sources and access strategies in Ugandan secondary cities: An intersectional analysis. *Environment and Urbanization*. Online First. 1–22. doi:10.1177/0956247819847346

Mackenzie, S. (1989). Women in the city. In: *New Models in Geography* (eds. R. Peet and N. Thrift), 109–226. London: Unwin Hyman.

Milazzo, A. and van de Walle, D. (2015). *Women left behind? Poverty and headship in Africa.* (Policy Research Working Papers). Washington, DC: World Bank.

Opiyo, P., Obange, N., Ogindo, H. et al. (2018). *The characteristics, extent and drivers of urban food poverty in Kisumu, Kenya.* (Consuming Urban Poverty Project Working Paper No. 4). University of Cape Town: African Centre for Cities.

Peake, L. (2016). The twenty-first century quest for feminism and the global urban. *International Journal of Urban and Regional Research* 40 (1): 219–227.

Peake, L. and Rieker, M. eds. (2013). *Rethinking Feminist Interventions into the Urban.* London: Routledge.

Riley, L. and Dodson, B. (2014). Gendered mobilities and food access in Blantyre, Malawi. *Urban Forum* 25 (2): 227–279.

Riley, L. and Hovorka, A. (2015). Gendering urban food strategies across multiple scales. In: *Cities and Agriculture – Developing Resilient Urban Food Systems* (eds. H. De Zeeuw and P. Drechsel), 336–357. London: Routledge.

Ruel, M. and Garrett, J. (1999). Overview of special issue on urban food security. *World Development* 27 (11): 1885–1889.

Skinner, C. (2018). Contributing and yet excluded? Informal food retail in African cities. In: *Urban Food Systems Governance and Poverty in African Cities* (eds. J. Battersby and V. Watson), 104–115. London: Routledge.

Smit, W. (2018). Current urban food governance and planning in Africa. In: *Urban Food Systems Governance and Poverty in African Cities* (eds. J. Battersby and V. Watson), 94–103. London: Routledge.

Stevano, S. (2019). The limits of instrumentalism: informal work and gendered cycles of food insecurity in Mozambique. *The Journal of Development Studies* 55 (1): 83–98.

Streetnet (2019). *Solidarity statement on Zimbabwe.* http://streetnet.org. za/2019/02/05/solidarity-statement-on-zimbabwe (accessed 5 February 2019)

Tacoli, C. (2012). *Urbanization, Gender and Urban Poverty: Paid Work and Unpaid Carework in the City.* London: International Institute for Environment and Development (IIED) and New York: United Nations Population Fund (UNFPA).

Tacoli, C. (2017). Food (in)security in rapidly urbanising, low-income contexts. *International Journal of Environmental Research and Public Health* 14 (12): 1554.

Tawodzera, G. and Chigumira, E. (2019). *Household food poverty in Epworth, Zimbabwe.* (Consuming Urban Poverty Project Working Paper No. 8). University of Cape Town: African Centre for Cities.

Tawodzera, G. and Chileshe, M. (2019). *Household food poverty in Kitwe, Zambia.* (Consuming Urban Poverty Project Working Paper No. 6). University of Cape Town: African Centre for Cities.

Tevera, D. and Simelane, N. (2016). Urban food insecurity and social protection. In: *Rapid Urbanisation, Urban Food Deserts and Food Security in Africa* (eds. J. Crush and J. Battersby), 157–168. Cham, CH: Springer.

Tinker, I. (1997). *Street Foods: Urban Food and Employment in Developing Countries.* Oxford: Oxford University Press.

United Nations (2015). *World urbanisation prospects: The 2014 revision* (*ST/ESA/SER.A/366*). New York: UN Department of Economic and Social Affairs, Population Division.

Winders, J. and Smith, B.E. (2019). Social reproduction and capitalist production: A genealogy of dominant imaginaries. *Progress in Human Geography* 43 (5): 871–889.

Young, G. and Crush, J. (2019). *Governing the informal food sector in cities of the Global South.* (Hungry Cities Partnership [HCP] Discussion Paper 30). http://hungrycities.net/publication/hcp-discussion-paper-no-30-governing-informal-food-sector-cities-global-south

11

Infrastructures of Social Reproduction

Dialogic Collaboration and Feminist Comparative Urbanism

Tom Gillespie (University of Manchester)
Kate Hardy (University of Leeds)

Introduction

Following calls to rethink Western-centric urban theory (Robinson 2006; Roy 2009), there is now a growing interest in comparison as a tool to build theory from the experiences of a greater diversity of cities (Robinson 2016a, 2016b). However, the historical divide between feminist urban scholarship on the global North and South remains largely unaddressed (Peake 2016a), and comparative urban research has neglected feminist analyses of the urban politics of gender (Binnie 2014). As such, both feminist urban scholarship and comparative urbanism can benefit from being brought into dialogue in order to generate theory that is informed by gendered urban struggles in diverse contexts.

We argue that feminist comparative urbanism can inform a theorization of the urban as constituted by everyday struggles over social reproduction that exceed capitalist processes of urbanization, yet which are central to the reproduction of urban life. We propose a method of feminist comparative urbanism in order to explore two seemingly disparate urban movements – sex worker union organizing in Córdoba, Argentina, and housing activism in London, United Kingdom (UK) – as struggles over

A Feminist Urban Theory for our Time: Rethinking Social Reproduction and the Urban, First Edition. Edited by Linda Peake, Elsa Koleth, Gökbörü Sarp Tanyildiz, Rajyashree N. Reddy & darren patrick/dp.

'infrastructures of social reproduction' (Ruddick et al. 2018, p. 396). Our engagement with these movements is informed by our methodology of 'dialogic collaboration' that privileges the situated, everyday knowledge of activists and, following Mohanty (2003), emphasizes practising solidarity. Dialogic collaboration is characterized by four elements: i) Situated knowledge; ii Solidarity; iii) Collaboration; and iv) Iteration. We combine this methodology with a comparative analysis to explore how these place-based struggles are both particular to their historical-geographical context and simultaneously constituted by, and constitutive of, shared processes that shape the everyday lives of sex workers and homeless single mothers.

In this chapter, after reviewing the debates around feminist urban scholarship and comparative urbanism, we introduce the two case study movements as well as our approach to comparison. We then turn to our method of dialogic collaboration to explain how this developed through our engagements with the two movements in question. In the following section, the cases are compared across three axes: subjectivation, demands, and strategy. Finally, we discuss the insights generated by this comparison into the urban politics of infrastructures of social reproduction, and the significance of these insights for conceptualizing the urban.

Feminist Urban Scholarship and Comparative Urbanism

Social reproduction has long been a key concept for feminist scholars, and materialist and Marxist-feminists in particular, but has experienced a recent renaissance across the social sciences (cf Bhattacharaya 2017). According to Brenner and Laslett (1991), the concept of social reproduction refers to the reproduction of life itself, in contrast to societal reproduction, which denotes the reproduction of capitalist social relations. In this sense, social reproduction is constituted by 'the material social practices through which people reproduce themselves on a daily and generational basis' (Katz 2001, p. 709).

The Marxist-feminists of the 1970s (Dalla Costa 1972; Dalla Costa and James 1972; Federici 1975) extended and reinvigorated Marxist theory, asserting the necessity of socially reproductive activities to capitalist production. Incorporating these critiques into urban theory, feminist scholars sought to denaturalize and problematize women's spatial experience of the Western industrial city, using the concept of social reproduction to show 'how women's unpaid labour in the family context contributed to day-to-day and generational reproduction' (Rose 2010, p. 395). These early contributions tended to conceptualize reproductive labour as 'housework' or 'domestic labour', located in the home (Dalla Costa and James 1975; Federici 1975; MacKenzie and Rose 1983), and largely distinct from 'production'. In contrast, later work by feminist geographers sought to 'diminish the pertinence of the production-reproduction divide' (Rose 2010, p. 125) and the separate spheres approach, insisting on their dialectical and interconnected nature (MacKenzie 1989). At the turn of the century, Mitchell et al. (2003) revived and reinvigorated theorizations of 'work and non-work', arguing for a qualitatively different understanding of them as deeply interpellated under neoliberal capitalism.

Despite this emphasis on the importance of reproductive practices to capitalist value creation, social reproduction is also an immanent terrain of struggle (Mitchell et al. 2015). Public libraries (Frederiksen 2015), energy infrastructures (Angel 2019), and community gardens (Engel-Di Mauro 2018) are just a few of the urban sites of social reproduction in which capitalist disinvestment and dispossession can be contested by collective struggle. Ruddick at al. (2018, p. 396) argue that urban social movements can be understood as mobilizations around access to 'infrastructures of social reproduction'. Drawing on Brenner and Laslett (1991) and Katz (2001), we understand these urban infrastructures of social reproduction as the socio-material resources, practices, and spaces that enable the reproduction of human life. These include, amongst others, water, sanitation, transport, care, and importantly for the two movements discussed here, housing, education, and healthcare.

The field of feminist urban studies has a history of division between scholarship on the global North and South, with a lack of dialogue between them (Peake 2016a). This remains the case despite calls from postcolonial urban scholars for a new approach to theory production that disrupts hierarchical distinctions between Northern and Southern cities (Robinson 2006, 2011; Roy 2009). Within this context, comparison has been identified as a key tool for questioning and rethinking established, Western-centric urban theory by bringing diverse contexts into dialogue. Robinson (2016b, p. 20) proposes a typology of comparative strategies and methodological tactics in order to explore both interconnections ('genetic' comparison) and shared features ('generative' comparison) between cases. In addition, interventions drawing on relational geographical approaches (Ward 2010) and Marxist dialectics (Hart 2018) advocate the importance of comparison for understanding the shared relationships and processes that connect cities. However, recent debates on comparative urbanism have largely neglected the urban politics of gender (Binnie 2014), despite the fact that feminist analyses have 'informed most of the central debates in urban geography' for over 50 years (Pratt 1990, p. 595).

We argue that bringing comparative urbanism and feminist urban scholarship into dialogue can enhance both traditions. Robinson (2011, p. 18) argues that a postcolonial approach to comparison can generate new urban theory that is grounded in 'nuanced, complex and contextual accounts of urban processes' that do not take Western cities as their exemplar. Feminist research methods can play a valuable role in achieving this ambition through the production of situated urban knowledge. Given that all knowledge is directly tied to its epistemic community (i.e. those who produce it), knowledge is always situated, specific, and partial to the location from which the knower 'knows' (Haraway 1988). In addition, materialist feminist research has historically sought to theorize from this situated knowledge in order to understand and transform the social totality, understood here as the interconnectedness and co-constitution of social relations in their entirety (Hennessy 1993; Weeks 1998). Comparative urbanism informs an understanding of how particular urban struggles are related through common 'constitutive processes' (Hart 2018, p. 375). As such, we argue that feminist comparative urbanism builds on, incorporates, and synthesizes Marxist, postcolonial, and relational approaches to urban theory production.

Thinking Comparatively Between Córdoba and London

In what follows, we explore the potential for feminist comparative urbanism through a study of two urban movements: sex worker union organizing in Córdoba and a housing campaign led by single mothers in London. Despite the significant differences between these movements and their historical-geographical contexts, we argue that they share certain features that render them comparable. Drawing on the work of Robinson (2016a, 2016b), we argue that comparing across difference in these cases can generate new insights into the urban politics of infrastructures of social reproduction.

The Asociación de Mujeres Meretrices de Argentina (AMMAR) is a sex workers' union that first emerged in Buenos Aires in 1994 as a collective response to police repression. Beginning with the demand to work freely, sex workers began to organize under the umbrella of the radical trade union federation, the Central de Trabajadores Argentinos (CTA) (Hardy 2010). AMMAR's chapter in Córdoba, Argentina's second biggest city, was established in the year 2000. Following political divergences, the Córdoba chapter broke away from the national organization in 2014, although they continue to organize under the umbrella of the CTA. While organizing as workers within a trade union, AMMAR should be understood as an organization engaged in reproductive struggle for two reasons. First, freedom from police violence is a reproductive justice issue (Rogers 2015). The frequent arrest and detention of sex workers prevented them from carrying out reproductive labour, including breastfeeding their young children. Second, following the initial demand of stopping police violence, AMMAR have increasingly organized around access to healthcare and education – key infrastructures of social reproduction – in their struggle to improve their lives.

Focus E15 is a housing campaign led by women in London. The UK capital is experiencing a severe housing crisis driven by a complex intersection of factors, including escalating housing costs, the long-term erosion of the city's social housing stock, and deep welfare cuts (Watt and Minton 2016). In this context, London's local authorities are increasingly managing their growing homeless populations by placing them in private temporary accommodation, often outside their own borough or the city altogether (Hardy and Gillespie 2016). The East London borough of Newham has sought to cut its waiting list for social housing by prioritizing applicants in paid employment and ex-members of the armed forces, resulting in the displacement of homeless female single parents (Watt 2018a). The Focus E15 campaign was established in Newham in 2013 after a group of homeless single mothers were issued with eviction notices from the Focus E15 hostel and told they would be rehoused up to 200 miles away, in cities such as Birmingham and Manchester. These women established the Focus E15 campaign to contest their displacement and demand access to social housing in London.

AMMAR and Focus E15 must be understood within their very different historical-geographical contexts. Economically, London is a 'global' or 'world' city with a dominant financial sector, whereas Córdoba is a middle-income city with large industrial and informal sectors. In addition, the extensiveness of the welfare state differs significantly between these two contexts. Welfare states have historically been constructed in order to decommodify social reproduction (Epsing-Anderson 1990), yet the degree of decommodification, and which social actors are included, differs between the UK and Argentina. The post-war welfare state in the UK was founded on universalist principles, grounded in the figure of the citizen (Pla and Ayos 2018), and included the extensive provision of social housing: 30.1% of all Londoners were council housing tenants in 1981 (Watt 2018b). However, this provision has subsequently been rolled back by neoliberal privatization policies, with council housing accommodating only 13.5% of Londoners in 2011 (ibid). Welfare retrenchment has been justified by a hegemonic 'anti-welfare common-sense' that stigmatizes welfare recipients as idle and parasitical (Jensen and Tyler 2015, p. 470). Austerity policies have intensely gendered effects (Hall 2018), with low-income single mothers particularly disadvantaged by welfare cuts (Watt 2018a). It is within this context that the women-led Focus E15 campaign emerged to demand access to a key infrastructure of social reproduction: housing (Gillespie, Hardy, and Watt 2018).

In contrast to the formation of the British welfare state, its Argentine equivalent can be considered a form of stratified universalism. Forged under President Juan Perón in the 1940s, this welfare state built on and systematized 19th-century social protection measures, characterized largely by state subsidies granted to philanthropic associations (Guy 2008). In this model, social insurance was, and largely remains, linked to occupation. According to Pla and Ayos (2018, p. 5), the Argentine welfare state was 'organised on the image of formal salaried workers', rather than on the figure of the citizen. Social insurance enabled such workers to access healthcare via trade unions, resulting in a stratified tripartite system of universal, but low-quality, public healthcare alongside higher-quality private and union-based provision. Due to the numerical dominance of unregistered workers in Argentina, high-quality healthcare has only been available to a minority of citizens. Moreover, despite the shift from neoliberalism to neo-developmentalism in the early 21st century, characterized by a renewed expansion of welfare policies, those employed outside the formal economy remained largely excluded (Hardy 2016). As a result, even accounting for the recent corrosion of universalism in the UK, welfare provision in Argentina has always been more limited and uneven.

It is important to recognize that these contextual differences are not coincidental; as decolonial scholars argue, the European colonization of Latin America has fundamentally shaped the modern world (Asher 2013). Following independence from Spain in 1810, Argentina played an important role as a 'subordinate partner' (Markham 2017, p. 450) within Britain's 'informal' empire (Gallagher and Robinson 1953) as both a source of primary commodities and an outlet for surplus capital through the construction of Argentina's ports and railways. It is within the context of this exploitative, quasi-colonial relationship that Argentina became an industrial society with a large urban working class, prompting Perón to extend welfare provision for the growing formal workforce (Romero 2001). Meanwhile, imperialist exploitation enabled the British state to finance the expansion of the welfare state in the early 20th century (Gough 1979). As such, the varied extensiveness of the welfare state in Córdoba and London reflects the different relationships of European and Latin American cities to historical processes of colonial exploitation.

Due to the contextual differences discussed above, traditional, quasi-scientific comparative methods would almost certainly consider Córdoba and London incomparable (Robinson 2011). However, we take inspiration from Robinson's (2006) call to understand all cities as 'ordinary', and therefore potentially comparable. Departing from the traditional comparative approach that seeks to limit variation between cases, Robinson (2016a, p. 194) advocates comparing across difference as the basis of 'conceptual innovation'. In addition, bringing the experiences of socially and economically marginalized Argentinean and British women into conversation in order to theorize the urban can contribute to decolonial knowledge production by destabilizing the Eurocentric paradigm of rationality that is premised on an epistemological separation of Western 'subjects' from non-Westerners who are stripped of agency and rendered as 'objects' (Quijano 2007).

Drawing on Robinson's (2016b, p. 16) typology of comparative strategies, we employ a 'generative' approach in which 'shared features provide the opportunity for thinking with the variety of the selected phenomena to generate conceptual insights.' The shared features of our cases include stigmatized female subjects organizing collectively around access to infrastructures of social reproduction: education and healthcare (Córdoba) and housing (London). Our understanding of these shared features, and therefore the comparability of these movements, is grounded in a social ontology of the urban as constituted through the everyday struggles of urban subjects (Ruddick et al. 2018). The following section explains how we conducted research with these two movements through developing a method of 'dialogic collaboration'.

Dialogic Collaboration

In Córdoba, Hardy conducted participant observation over four fieldwork visits (2007, 2008, 2014, 2019). During these visits, she conducted 32 interviews (17 with AMMAR members/activists, 9 with non-union sex workers, and 6 with other actors) and 41 questionnaire surveys with unionized and non-unionized sex workers. This is a sub-set of a wider project at the national level in which 120 interviews were undertaken as well as 297 questionnaires (see Hardy 2010, 2016). In London, between 2014 and 2016, Gillespie and Hardy conducted auto-ethnography through participation in the Focus E15 campaign. We also conducted participatory action research (PAR) on homelessness and displacement in East London in collaboration with the campaign. Sixty-four structured interviews were conducted with people experiencing housing insecurity, eliciting both quantitative and more qualitative data on the embodied, gendered experience of London's housing crisis (see Hardy and Gillespie 2016). In addition, Gillespie, Hardy, and Paul Watt conducted 12 interviews with campaign members and others involved in the 2014 occupation of a public housing estate in Newham (see Gillespie, Hardy, and Watt 2018).

In both Córdoba and London, these data were generated through a method that we call 'dialogic collaboration'. This method is influenced by feminist standpoint theory, Marx's social ontology (Ruddick et al. 2018), and participatory methods developed in Latin America (Freire 1970; Winton 2007). It is characterized by four elements: i) Situated knowledge; ii) Solidarity; iii) Collaboration; and iv) Iteration. Participants in both cases were given the opportunity to anonymize their participation in the research. However, many opted to use their real, legal names, seeking to visibilize their activism and labour. In the case of AMMAR, using real names has been an explicit part of activists' struggle against stigmatization.

Situated Knowledge

Dialogic collaboration grants epistemic privilege to actors engaged in everyday struggle, and involves those actors in the production of knowledge about themselves. Rather than theorizing 'everything from nowhere' (Haraway 1988, p. 581), in what follows we begin by theorizing from the embodied knowledge of Córdoban sex workers and London's homeless single mothers. The aim is to produce specific, localized knowledge in order to generate understandings of the 'social totality' (Goonewardena 2018, p. 466) in and through which these struggles are constituted. Such knowledge can contribute to the materialist feminist project to mobilize social reproduction theory in service of understanding capitalist relations as a totality (Weeks 1998; Ferguson et al. 2016).

Situated knowledge requires that researchers reflect on the constitutive role of their own positionality in shaping the research process. Our engagement as researchers in both sites was ethnographic and autoethnographic and characterized by differing degrees and types of involvement. Argentina was selected as AMMAR represents one of the most active sex workers' unions in the world. As such, Hardy's relationship with AMMAR was developed specifically for the purposes of research, albeit building on ongoing activist work with sex worker organizations in the UK. In 2007, Hardy travelled to Buenos Aires and met with Elena Reynaga, the General Secretary of AMMAR, with whom she had previously communicated over email. She met six regional General Secretaries and met with a heavily pregnant María-Eugenia Aravena, General Secretary of AMMAR-Córdoba. Aravena was keen to participate and encouraged Hardy to return the following year for a longer period of time. As well as interviews, Hardy went on *recorridas* (outreach walks around working areas), distributing condoms and talking to workers. While Hardy's identity as a white, middle-class, Western European woman identified her as racially distinct and relatively privileged, this did not appear to act as a barrier to building relationships. The CTA (and therefore AMMAR) conceptualize all who rely on the wage (including university researchers) as members of the 'working class', making Hardy more comprehensible as an ally (see Hardy 2012). Following another solo visit by Hardy in 2014, both Gillespie and Hardy spent a week with AMMAR-Cordoba in 2019. This involved sharing meals and yerba mate tea with activists at the union headquarters and in their homes.

In contrast, both authors were actively involved in Focus E15 as campaign members between 2014 and 2016 (see Gillespie, Hardy, and Watt 2018). In 2014, Hardy approached Focus E15 as part of the London-based feminist collective 'Feminist Fightback', while Gillespie encountered them through his participation in a local tenants' organization. Both authors began to attend the Focus E15's weekly stall and became increasingly involved through activities which included: promoting the campaign on social media; creating the campaign website; providing eviction support for a local family who had contacted the campaign; and helping with fundraising activities. During the 2014 occupation of empty public housing units, we drove a van to deliver furniture and other resources to the occupied flats, wrote press releases and contacted journalists, and participated in the daily life of the occupation. During Focus E15's 'march against evictions' in 2015, we spent several hours inflating hundreds of helium balloons that were subsequently released into the sky to represent those who have been displaced from London as a result of the housing crisis. Despite this insider role as campaign members, however, our identities as middle-class academics, one of us a man, neither of us parents at the time, meant that we differed from the working-class mothers who founded and led the campaign. Yet, the universal demands of the Focus E15 campaign means that it has become a cross-class alliance, including a range of people from different backgrounds. Research processes are constitutive of researchers' subjectivities, and both authors credit much of our present political orientations to our engagement with these two movements.

Solidarity

Beyond tokenistic 'performances of candor' (Chakravarty 2015, p. 25), dialogic collaboration foregrounds situated knowledge as a basis for transformative politics. As such, an ethic of feminist solidarity undergirds dialogic collaboration through a commitment to producing knowledge that is relevant and useful to movement actors. In both cases, we sought to take responsibility as academics and allies well beyond the ethnographic moment and develop long-term collaborative relationships. Non-academic research reports were produced for both campaigns in order to 'co/produc[e] knowledges that "speak" the theoretical and political languages of communities' (Nagar 2003, p. 65). Between 2014 and 2016, Gillespie and Hardy contributed various forms of labour as members of the Focus E15 campaign, as outlined above. Despite leaving London, both authors have continued to support the campaign in various ways. For example, the participatory action research project enabled Focus E15 to collect data to inform a campaign around housing and mental health (Hardy and Gillespie 2016). Additionally, this research led to the foundation of a nationwide Housing and Mental Health Network, enabling Focus E15 to work with psychologists to provide mental health support for people facing housing insecurity and displacement. Reflecting on the project, the Focus E15 steering committee said 'the research project has increased capacity within the campaign … building this bank of experience has been a very useful resource for ourselves as activists in how we can support other individuals facing homelessness and how to respond to the council's actions' (personal communication 2017). In addition, we have both continued to attend occasional meetings and demonstrations, promoting the campaign through social media and academic work, as well participating in social events such as birthday parties.

Hardy's relationship with AMMAR has been maintained through regular visits over the last decade (2007, 2008, 2014, 2019); regular, sometimes daily, email and WhatsApp contact; undertaking fundraising; and through using her own time to identify funds and co-write funding proposals with the aim of supporting the organization. In June 2019, Hardy wrote an article about Argentine sex worker organizing for *The Guardian* newspaper in order to help raise the international profile of AMMAR and donated her fee to the union (Hardy 2019). These sustained solidarities seek to put into practice arguments that comparative urban research should be grounded in a long-term commitment to the contexts studied (Robinson 2011) and contribute to praxis beyond the academy (Hart 2018).

Collaboration

Collaborative methods were employed in order to privilege participants' perspectives. By centering those subjects about whom knowledge is being created from the beginning of the research process, collaboration can make research 'useful' and 'relevant' beyond the academy (Staeheli and Nagar 2002). Hardy worked directly with the leaders of AMMAR to develop research questions that addressed their concerns about why some women chose to participate in the movement while others did not. This generated data that enabled union leaders to comprehend (and seek to overcome) obstacles to organizing that individual women faced, such as low self-esteem, family commitments, and stigma (see Hardy and Cruz 2018). While Hardy's relationship with AMMAR began as a research collaboration, 'our involvement in the [Focus E15] campaign preceded any intention to write about it' (Gillespie, Hardy, and Watt 2018, p. 5). Through participating in the campaign's weekly street stall, we began to notice that local residents were approaching campaigners with their own stories of homelessness and displacement. Following a suggestion by the authors to collect this information as data, and through subsequent discussions in campaign meetings, it was decided by the campaign steering committee that the authors would collaborate with campaigners to design an action research project in order to capture these experiences (see Hardy and Gillespie 2016).

Iteration

Dialogic collaboration emerges iteratively through an ongoing dialogue between researchers and participants. In Córdoba, data collection methods were designed through a process of negotiation. Influenced by the notion of participatory design (Freire 1970) as a method for emancipatory research, Hardy initially travelled to Argentina with a proposal for using 'participatory mapping', a method used successfully by other scholars in Latin America (Winton 2007). Yet during a meeting in which Hardy proposed the use of mapping, a union leader flatly asked: 'you're the researcher, why don't you do the research?' – indicating a preference for more traditional methods. AMMAR leaders felt that a questionnaire-based survey would generate quantitative data that would be more useful for negotiating with state actors. In addition, this structured method enabled sex workers with varying degrees of education and literacy to be trained and paid as peer researchers, using the questionnaire tool designed by Hardy to collect data from their *compañeras*.

Following the decision to undertake PAR in London, Gillespie and Hardy met with the Focus E15 steering committee to identify research questions and develop appropriate methods. As with AMMAR, the committee selected a questionnaire survey, both to facilitate ease for peer researchers drawn from the campaign and to generate quantitative data that would build legitimacy when engaging with authorities. The authors developed a questionnaire tool and piloted it with campaign members. The tool went through several iterations, drawing on campaigners' own experiences of housing insecurity to refine the questions. The steering committee identified which campaigners would be trained and paid as peer researchers. These researchers then utilized their embodied knowledge of homelessness and displacement to identify potential participants at Newham Council's housing office and at various temporary accommodation locations in and around London.

This method of 'dialogic collaboration' was developed initially in Córdoba before travelling to London. As such, these cases were initially related through the transnational circulation of research methods. The trajectory of this research method with a circulatory dynamic between South and North responds to calls for multi-directional circuits of knowledge production (Robinson 2011) and 'learning to learn from below' (Spivak 2000, p. 327) and more specifically, in collaboration across these contexts.

Gendered Urban Struggles in Córdoba and London

Dialogic collaboration incorporates standpoint theory's conceptualization of the subject, asserting that it is women's (and others') social position that shapes their subjectivity (Hennessy 1993). Theorizing knowledge as emerging from struggle and foregrounding lived experience, our analysis in this section begins from a discussion of the subjectivities constitutive of, and constituted by, the urban politics of infrastructures of social reproduction. Next, we analyse the content of the demands that emerged following the subjectivation of these actors through struggle, as they shift from victimized, stigmatized, and invisibilized subjects to agential actors with collective strategies for changing the conditions in which they live (see Hardy and Cruz 2018 for more on this in relation to AMMAR). Finally, we discuss the diverse strategies employed to address these demands.

Subjectivation

In both cases, the movements were either led by women (Focus E15) or populated solely by women who are marked by significant social stigma (Morcillo 2014). The stigma faced by both groups is premised both on sexual mores and also narratives which render single mothers and sex workers as unproductive or 'lazy'. Raquel (AMMAR member) explained that while 'everyone calls [prostitution] "la vida fácil" ("the easy life") … it is not easy for anyone.' Similarly, Jasmin (Focus E15) explained that when they began their campaign, 'we had a lot of people approaching us randomly saying things about us being on benefits, taking from the system, things like that.'

In the face of such stigma and with little political experience, both movements initially mobilized around notions of motherhood. As Eugenía (Chair of AMMAR-Córdoba) argued, the trope of motherhood was important for:

delegitimating the morbid fascination that society has [with sex work by saying] that the women are mothers, family, they maintain their homes, that they started working out of economic necessity, that it's a social problem.

This approach echoes that of sex worker rights movements in France and England, who have mobilized the identity of motherhood to elicit support (Jaget 1980). Furthermore, sex workers in Córdoba could also draw on the rich history of political motherhood evident in the repertoires and narratives of Argentinian social movements. The Madres de la Plaza de Mayo organized together in the 1970s and 1980s to demand the safe return of their children who had been 'disappeared' by the military dictatorship. Their identities as mothers chimed with the 'family values' morality of the junta, affording them the political space to publicly challenge the authorities in an otherwise repressive climate (Fisher 1993). Despite initially drawing on this history, however, engaging with union activists in the CTA led AMMAR to increasingly eschew their identities as mothers through undergoing a process of class-based subjectivation grounded in identities as sex workers (Hardy and Cruz 2018). Subjectivation can be understood as 'a complex set of processes whereby one's sense of self as an individual agent is paradoxically shaped according to processes and forces external to the self' (Langlois and Elmer 2019, p. 237). Eugenía stated that AMMAR should 'stop apologising' and saying that their engagement in sex work was because 'I have five children, I can't read or write, I'm poor.' Rather, she argued that the women should demand recognition on the basis of their position as members of the working class.

In contrast to the positive associations with motherhood in Argentinian politics, working-class single mothers in the UK have routinely been demonized by anti-welfare discourses as a drain on public resources (Tyler 2008; Cain 2015). As such, the identity of single motherhood is an unlikely basis for making political demands in this context. Despite this, Focus E15 received a largely sympathetic media and public response to their campaign. As campaigner Emer argues, images of homeless women with children resonated with popular concern about the deepening housing crisis in London:

Mothers having children who just want to have a roof over their head for their child to be secure ... It's something that gets people ... [it's] a narrative that people can understand and get behind.

Despite emerging from the particular experiences of the mothers, the campaign has expanded over time to address the needs of others facing housing insecurity, including men. As such, although the campaign was initially named the Focus E15 Mothers, it later became known as the Focus E15 campaign. After encountering other housing campaigns through London's Radical Housing Network, this name change reflected a growing appreciation that the mothers' experiences were not exceptional. Rather, they were symptomatic of a much wider housing crisis that had not only gendered, but also classed and raced dimensions. As such, the campaign's initial focus on the mothers' own housing situation broadened out into a demand for social housing for all. According to campaign member Andrew, although

> [Focus E15] was started by mothers in a hostel, they decided to link up, so they're not [only] doing it for themselves. It's not a kind of economistic demand: 'We want this for ourselves.' It's: 'We want this for other people as well.'

In both cases, women experiencing stigmatization, oppression, and exclusion underwent a process of subjectivation that occurred through encounters with other contentious actors. While motherhood was initially mobilized by AMMAR and Focus E15, therefore, both movements have moved beyond this particular identity in order to demand access to infrastructures of social reproduction, notably housing, education, and healthcare.

Demands

In both Córdoba and London, this subjectivation of women as workers and housing rights campaigners led to the emergence of new demands. In the case of AMMAR, the focus shifted away from police repression to a broader consideration of the multidimensional needs of sex workers:

> As time passed, we began to see that we also had needs. Not only to fight against unjust police repression, but also to be able to work with issues around health, education and others (Eugenía).

Street sex workers in Argentina are characterized by low educational levels and poor access to healthcare. Eleven percent of respondents to Hardy's survey had completed no education, and 63% had primary level education only. Only a minority (25%) of women had completed secondary school (Hardy 2016). Despite the availability of public healthcare in principle, the system of queuing for an appointment limits access for sex workers in practice, who often have to choose between working and waiting in line. Moreover, discrimination by medical staff obfuscates access to appropriate healthcare for sex workers. As a result, and drawing on Argentina's historic politics of subsidies, AMMAR have developed a series of demands directed at the local state in order to enable access to infrastructures of reproduction. Unlike formal political demands, these tend to take the form of requests for officials to address members' healthcare and educational needs. Raquel recounted how AMMAR enabled access to healthcare for women by demanding fair treatment within existing structures: 'We told [medical staff] they had to treat them well, not to discriminate against them, not to treat them like dogs, because this is what [the women] were afraid of.'

The shifting identity of Focus E15 from a mothers' campaign to a broad-based housing movement was also reflected in changing demands over time. Initially the women simply petitioned Newham Council to rehouse them in the borough. As the focus of the campaign shifted to the wider issue of housing justice in London, they adopted the broader demand for 'social housing, not social cleansing'. To this end, Focus E15 has sought to draw attention to the extreme contradictions of London's housing crisis, such as empty housing in the midst of homelessness, by demanding that Newham 'repopulate the Carpenters Estate'. The Carpenters is a post-war public housing estate that the Council is currently emptying with a view to redeveloping as part of its strategy to gentrify the borough. In September 2014, Focus E15 occupied two empty units on the estate for two weeks in order to demand that the Council open up much-needed social homes to the homeless (Gillespie, Hardy, and Watt 2018). Jasmin, a founder of the campaign, understands this as part of a wider struggle against austerity:

> It's not just about housing. It's about cuts to the NHS, it's about schools being shut, community centres being shut down, privatisation, every-thing. And housing is a big part of that. It's the first thing before anything else, I think.

Both AMMAR and Focus E15 emerged to resist the repression and displacement of stigmatized female subjects. In each case, however, the campaigns have moved beyond simply resisting state oppression to making demands on the state for access to infrastructures of social reproduction: education and healthcare (AMMAR), and housing (Focus E15).

Strategy

Both movements have also pursued a combination of different strategies in order to gain access to infrastructures of social reproduction in relation to education, healthcare, and housing. The first strategy, employed to differing degrees in both Córdoba and London, is to demand access to existing infrastructures that are produced and controlled by the state. Both AMMAR and Focus E15 have made claims on the state to expand their provision to excluded and marginalized subjects. Focus E15's key demand – 'social housing, not social cleansing' – is aimed specifically at the local state, which has historically been responsible for social housing provision in the UK. Sex workers in Córdoba have also engaged the local state at the provincial and city scale in order to challenge their traditional exclusion from public education and healthcare provision. They have done so by cultivating strategic relationships with particular public servants in local institutions. For example, AMMAR arranged special hospital appointments at hours suitable for sex workers through establishing a relationship with the local hospital's Director of Infectology.

Although both campaigns make demands of the local state for access to existing infrastructures of social reproduction, they also seek to create autonomous infrastructures through their everyday struggles. When Focus E15 occupied the Carpenters Estate, people visited from across London and beyond, participating in the creation of a temporary 'urban commons' characterized by the collectivization of reproductive activities (Gillespie, Hardy, and Watt 2018). For example, dedicated activities were programmed for children and participants took it in turn to provide childcare and cook collective meals. Although the primary aim of the occupation was to compel the state to rehouse people on the estate, a significant outcome was to create an autonomous space of social reproduction. According to campaign member Andrew: 'We take on the council, and we hate them, we fight them [but] we also create other things on the way.'

In Argentina, *Planes Jefes y Jefas de Hogares Desempleados* (Heads of Unemployed Households Plans) were introduced in April 2002, as part of a broader programme for social inclusion. The *planes* offered cash transfers in exchange for working at health centres, schools, and communal kitchens. By 2007, they supported over 2 million people (Grugel and Riggirozzi 2007). However, many sex workers have not been able to access these benefits due to an inability to navigate the application process (Hardy 2016). AMMAR ran workshops to support women to apply for the planes, but only a minority were successful. Excluded from state welfare provision, therefore, AMMAR has created 'autonomous sites for enacting … mutual care' (Hardy 2016, p. 99), or autonomous infrastructures of social reproduction. In Córdoba, a *compañera* with six children and no partner was unable to work for three months due to illness. In the absence of state support, AMMAR members made and raffled a cake to raise money for their sick comrade. Most strikingly, when a compañera was shot in a neighbourhood dispute, members were called upon to donate blood, as the hospital was in short supply. The union therefore created a space for sex workers to sustain each other with the most basic corporeal material for the biological reproduction of human life.

A third key strategy adopted by AMMAR is the co-production of infrastructures of social reproduction with state actors. AMMAR's headquarters, located on the edge of the city in the middle of the sex working district, is part funded by a contribution towards rent costs from the municipality. The unprepossessing white building is two storeys high. Downstairs contains a kitchen with a large table, around which women and other members of the community gather all day to smoke, swap stories, and eat a communal lunch. Every now and then, an adjacent door opens and a cacophony of sound rings out from the Sala Cuna: the first nursery in the country run by sex workers. The Province of Córdoba funds the Sala Cuna programme in order to enable female social and labour market inclusion. Upstairs, the door to the left is the entrance to the Centro Amigable de Salud (Friendly Health Centre), a collaboration between AMMAR and a gynaecologist from the local public hospital. The Centre offers gynaecologist appointments, sexual health testing, and support for drug use. Across the hallway, a row of small desks all face towards a blackboard in the primary school room. Classes begin at 2 pm, initially designed to suit the timetables of women who have been working all night, but the school now provides lessons to a range of working-class people. The municipality funds the teacher's wages and provides bus passes and hot meals in order to enable low-income women to attend. While state funding supports various elements of this multifaceted infrastructure of social reproduction, the building is managed by AMMAR, and union activists provide unpaid labour by undertaking cleaning and administrative duties.

In contrast to these examples of co-production in Córdoba, the relationship between Focus E15 and the local state has been largely antagonistic. The campaign has sought to publicize the mistreatment of homeless people by Newham Council, particularly under the leadership of former Mayor, Robin Wales. However, while Focus E15 has sought to responsibilize the state for the provision of social housing, it is noteworthy that the slogan 'we decorate, you populate' emerged during the Carpenters Estate occupation. While the campaign called on the Council to reinstate people in social homes on the Estate, Focus E15 offered to donate their own labour to make these homes habitable. Although the entire estate was not repopulated as a result of the occupation, the Council did agree to rehouse a small number of people on a temporary basis (Gillespie, Hardy, and Watt 2018). To a limited extent, therefore, this indicates a willingness amongst activists to work alongside the state to co-produce infrastructures of social reproduction.

Infrastructures of Social Reproduction and the Urban

Comparing these movements across three axes – subjectivation, demands, and strategy – demonstrates how comparative urbanism can draw on feminist methods of dialogic collaboration to generate new insights into differentially located struggles over infrastructures of social reproduction. Employing Robinson's (2016b, p. 22) strategy of 'generative' comparison, we identify several shared features between AMMAR and Focus E15: leadership by stigmatized women excluded from access to state-provided infrastructures; initial contestation through the strategic mobilization of motherhood; subjectivation through struggle; and a transition from resisting state violence to demanding access to infrastructures of social reproduction. Identifying these common 'constitutive processes' (Hart 2018, p. 375) demonstrates how feminist comparative urbanism enables partial theoretical insights into the totality of social relations which shapes both contexts, such as gendered dynamics of stigmatization and exclusion, and the agency of marginalized women in contesting these dynamics through collective struggle.

In addition to identifying shared features, generative comparison requires thinking through the variety between cases (Robinson 2016b). Both cases demonstrate how urban social movements combine strategies of 'contention' (demands to access existing infrastructures), 'subversion' (creating autonomous infrastructures), and 'collaboration' (co-producing new infrastructures with the state) in order to achieve change (Mitlin 2018, p. 557). However, it is possible to identify diversity between the two movements in the emphasis placed on each strategy. In particular, while AMMAR has willingly co-produced health and education services with the local state, Focus E15 has primarily focused on contentious demands for the redistribution of Newham's existing social housing stock. This divergence can be explained, in part, by the historical orientation of the state towards social reproduction in Argentina and the UK, as institutionalized in the extensiveness of the welfare state in both contexts. Despite decades of welfare retrenchment, Focus E15's demand for social housing demonstrates a residual expectation amongst Londoners that the state is ultimately responsible for providing infrastructures of social reproduction for all citizens. By contrast, AMMAR's co-production of education and healthcare infrastructures reflects an expectation amongst Córdobans that informal workers' movements must address gaps in the historically uneven welfare state – with its privileging of formal workers – with subsidized collective self-help. While there are common features that connect these two reproductive struggles, therefore, the strategies adopted by movement actors are specific to their historical-geographical contexts.

Comparing these movements adds empirical weight to Ruddick et al.'s (2018, p. 390) call for a social ontology of the urban as constituted by 'subjects, subjectivity and struggle'. In contrast to approaches that theorize the urban in terms of an all-encompassing capitalist process of urbanization, Ruddick et al. (2018, p. 399) instead understand urbanization as 'an open process determined through praxis, by actual people making the world they inhabit.' Considered together, AMMAR and Focus E15 demonstrate how the urban is produced through everyday struggles over infrastructures of social reproduction, and how women's subjectivities (as mothers, workers, and activists) are formed and reformed in the process. As such, these struggles can be understood as a 'constitutive outside' that exceed capitalist processes of urbanization but are central to the reproduction of urban life (Peake 2016a; Roy 2016; Reddy 2018). Furthermore, comparing these two movements illustrates the fundamental importance of historical difference in the constitution of the urban (Roy 2016). Depending on the historical-geographical context, struggles over infrastructures of social reproduction may take the form of the commoning of existing urban infrastructures, such as a council estate in London, or the creation of new infrastructures, such as a health centre and school in Córdoba. As such, feminist comparative urbanism enables a theorization of the urban as constituted by multiple everyday struggles over social reproduction that are both connected by common processes and distinguished by contextual differences.

Conclusion

Comparing sex worker union organizing in Córdoba and housing activism in London enables an exploration of the urban politics of infrastructures of social reproduction in two very different contexts. To this end, we propose a method of feminist comparative urbanism that aims to disrupt traditional epistemological divisions between Northern and Southern cities and is grounded in dialogic collaboration with social movements. Theoretically, this method can generate situated knowledge of the urban as constituted by everyday struggles over the means of social reproduction. Normatively, this method can inform feminist praxis by posing questions about what actions can be taken to create, demand, and defend infrastructures of social reproduction in an increasingly urbanized world. Although the cases of Córdoba and London enable comparison across differentially located struggles over infrastructures of social reproduction, these cities should not be understood as somehow representative of 'the North' and 'the South'. Rather, there is a need for further feminist comparative urban research that explores struggles over infrastructures of social reproduction in a greater diversity of contexts, including those with much less established welfare states than the cases discussed in this chapter. This research will enable comparative urbanism to play a valuable role in the production of a feminist urban theory for our time.

Acknowledgements

We would like to thank a range of people for helping produce this article. First and foremost, the inspirational activists of AMMAR and Focus E15. This research would not have been possible without funding support from the Feminist Review Fund, the University of Leeds, and the ESRC. We would like to thank Paul Watt, Emilio Ayos, Jessica Plá, Linda Peake, Gökbörü Sarp Tanyildiz, and one anonymous reviewer for helping us to develop the intellectual ideas discussed here. Finally, the chapter simply could not have been produced without the reproductive labour of Rosa Methol, Peggy Gillespie, Rowan Williams, Craig Griffiths, and the patience of Gabriel Gillespie Hardy.

References

Asher, K. (2013). Latin American decolonial thought, or making the subaltern speak. *Geography Compass* 7 (12): 832–842.

Bhattacharaya, T. ed. (2017). *Social Reproduction Theory: Remapping Class, Recentering Oppression*. London: Pluto Press.

Binnie, J. (2014). Relational comparison, queer urbanism and worlding cities. *Geography Compass* 8 (8): 590–599.

Brenner, J. and Laslett, B. (1991). Gender, social reproduction and women's organization: Considering the US welfare state. *Gender and Society* 5 (3): 311–333.

Cain, R. (2015). Responsibilising recovery: Lone and low-paid parents, universal credit and the gendered contradictions of UK welfare reform. *British Politics* 11 (4): 488–507.

Chakravarty, D. (2015). On being and providing 'data': Politics of transnational feminist collaboration and academic division of labor. *Frontiers: A Journal of Women Studies* 36 (3): 25–50.

Dalla Costa, M. (1972). Women and the subversion of the community. In: *The Power of Women and the Subversion of the Community* (eds. M. Dalla Costa and S. James), 19–54. Bristol: Falling Wall.

Engel-Di Mauro, S. (2018). Urban community gardens, commons, and social reproduction: Revisiting Silvia Federici's *Revolution at Point Zero*. *Gender, Place and Culture* doi:10.1080/0966369X.2018.1450731

Epsing-Anderson, G. (1990). *The Three Worlds of Welfare Capitalism*. Princeton, NJ: Princeton University Press.

Federici, S. (1975). *Wages against Housework*. Bristol: Falling Wall.

Ferguson, S., LeBaron, G., Dimitrakaki, A. et al. (2016). Special issue on social reproduction. *Historical Materialism* 24 (2): 25–37.

Fisher, J. (1993). *Out of the Shadows: Women, Resistance and Politics in South America*. London: Latin American Bureau.

Frederiksen, L. (2015). 'Our public library': Social reproduction and urban public space in Toronto. *Women's Studies International Forum* 48: 141–153.

Gallagher, J. and Robinson, R. (1953). The imperialism of free trade. *The Economic History Review* 6 (1): 1–15.

Gardiner, M. (2004). Everyday utopianism: Lefebvre and his critics. *Cultural Studies* 18 (2–3): 228–254.

Gillespie, T., Hardy, K., and Watt, P. (2018). Austerity urbanism and olympic counter-legacies: Gendering, defending and expanding the urban commons in East London. *Environment and Planning D: Society and Space* 36 (5): 812–830.

Goonewardena, K. (2018). Planetary urbanization and totality. *Environment and Planning D: Society and Space* 36 (3): 456–473.

Gough, I. (1979). *The Political Economy of the Welfare State*. London: Macmillan.

Grugel, J. and Riggirozzi, M.P. (2007). The return of the state in Argentina. *International Affairs* 83 (1): 87–107.

Guy, D. (2008). *Women Build the Welfare State: Performing Charity and Creating Rights in Argentina, 1880–1955*. Durham, NC: Duke University Press.

Hall, S.M. (2018). The personal is political: Feminist geographies of/in austerity. *Geoforum* doi:10.1016/j.geoforum.2018.04.010

Haraway, D. (1988). Situated knowledges: The science question in feminism and the privilege of partial perspective. *Feminist Studies* 14 (3): 575–599.

Hardy, K. (2010). Incorporating sex workers into the Argentine labor movement. *International Labor and Working-Class History* 77 (1): 89–108.

Hardy, K. (2012). Making space for emotion in development methodologies. *Emotion, Space and Society* 5 (2): 113–121.

Hardy, K. (2016). Uneven divestment of the state: Social reproduction and sex work in neo-developmentalist Argentina. *Globalizations* 13 (6): 876–889.

Hardy, K. (2019). Why UK feminists should embrace sex worker rights. *The Guardian* (3 June) www.theguardian.com/commentisfree/2019/jun/03/uk-feminists-embrace-sex-workers-rights (accessed 1 August 2019).

Hardy, K. and Cruz, K. (2018). Affective organizing: Collectivizing informal sex workers in an intimate union. *American Behavioral Scientist* doi:10.1177/0002764218794795

Hardy, K. and Gillespie, T. (2016). *Homelessness, health and housing: Participatory action research in East London*. Unpublished report.

Hart, G. (2018). Relational comparison revisited: Marxist postcolonial geographies in practice. *Progress in Human Geography* 42 (3): 371–394.

Hennessy, R. (1993). *Materialist Feminism and the Politics of Discourse*. London: Routledge.

Jaget, C. (1980). *Prostitutes: Our Life*. London: Falling Wall.

Jensen, T. and Tyler, I. (2015). Benefits broods: The cultural and political crafting of anti-welfare commonsense. *Critical Social Policy* 35 (4): 470–491.

MacKenzie, S. and Rose, D. (1983). Industrial change, the domestic economy and home life. In: *Redundant Spaces in Cities and Regions* (eds. J. Anderson, S. Duncan, and R. Hudson), 155–200. London: Academic Press.

Markham, B. (2017). The challenge to 'informal' empire: Argentina, Chile and British policy-makers in the immediate aftermath of the First World War. *The Journal of Imperial and Commonwealth History* 45 (3): 449–474.

Mitchell, K., Marston, S., and Katz, C. (2004). *Life's Work: Geographies of Social Reproduction*. Oxford: Wiley-Blackwell.

Mitlin, D. (2018). Beyond contention: Urban social movements and their multiple approaches to secure transformation. *Environment and Urbanization* 30 (2): 557–574.

Mohanty, C. (2003). *Feminism without Borders: Decolonizing Theory, Practicing Solidarity*. Durham, NC: Duke University Press.

Morcillo, S. (2014). 'Como un trabajo' Tensiones entre sentidos de lo laboral y la sexualidad en mujeres que hacen sexo comercial en Argentina. *Sexualidad, Salud y Socieded* 18: 12–40.

Nagar, R. with Ali, F. (2003). Collaboration across borders: Moving beyond positionality. *Singapore Journal of Tropical Geography* 24 (3): 356–372.

Peake, L. (2005). The Suzanne Mackenzie memorial lecture: Rethinking the politics of feminist knowledge production in geography. *The Canadian Geographer* 59 (3): 257–266.

Peake, L. (2016a). The twenty-first century quest for feminism and the global urban. *International Journal of Urban and Regional Research* 40 (1): 219–227.

Peake, L. (2016b). On feminism and feminist allies in knowledge production in urban geography. *Urban Geography* 37 (6): 830–838.

Pla, J.L. and Ayos, E.J. (2018). Producción de Bienestar y Estructura Social en Perspectiva Comparadad: Reino Unido, España y Argentina. *Ciudadanías. Revista de Políticas Sociales Urbanas* 3 (2do semestre): 106–134.

Pratt, G. (1990). Feminist analyses of the restructuring of urban life. *Urban Geography* 11 (6): 594–605.

Quijano, A. (2007). Coloniality and modernity/rationality. *Cultural Studies* 21 (2–3): 168–178.

Reddy, R.N. (2018). The urban under erasure: Towards a postcolonial critique of planetary urbanization. *Environment and Planning D: Society and Space* 36 (3): 529–539.

Robinson, J. (2006). *Ordinary Cities: Between Modernity and Development*. London: Routledge.

Robinson, J. (2011). Cities in a world of cities: The comparative gesture. *International Journal of Urban and Regional Research* 35 (1): 1–23.

Robinson, J. (2016a). Comparative urbanism: New geographies and cultures of theorizing the urban. *International Journal of Urban and Regional Research* 40 (1): 187–199.

Robinson, J. (2016b). Thinking cities through elsewhere: Comparative tactics for a more global urban studies. *Progress in Human Geography* 40 (1): 3–29.

Rogers, A. (2015). How police brutality harms mothers: Linking police violence to the reproductive justice movement. *Hastings Race & Poverty Law Journal* 12 (2): 205–234.

Roy, A. (2009). The 21st-century metropolis: New geographies of theory. *Regional Studies* 43 (6): 819–830.

Roy, A. (2016). What is urban about critical urban theory? *Urban Geography* 37 (6): 810–823.

Ruddick, S., Peake, L., Tanyildiz, G.S. et al. (2018). Planetary urbanization: An urban theory for our time? *Environment and Planning D: Society and Space* 36 (3): 387–404.

Spivak, G.C. (2000). Discussion: An afterword on the new subaltern. In: *Subaltern Studies XI: Community, Gender and Violence* (eds. P. Chatterjee and P. Jeganathan), 305–334. New Delhi: Permanent Black.

Staeheli, L. and Nagar, R. (2002). Feminists talking across worlds. *Gender, Place and Culture* 9 (2): 167–172.

Sundberg, J. (2003). Masculinist epistemologies and the politics of fieldwork in Latin Americanist geography. *The Professional Geographer* 55 (2): 180–190.

Tyler, I. (2008). 'Chav mum chav scum' class disgust in contemporary Britain. *Feminist Media Studies* 8 (1): 17–34.

Usami, K. (2004). Transformation and continuity of the Argentine welfare state: Evaluating social security reform in the 1990s. *The Developing Economies* 42 (2): 217–240.

Ward, K. (2010). Towards a relational comparative approach to the study of cities. *Progress in Human Geography* 34 (4): 471–487.

Watt, P. (2018a). Gendering the right to housing in the city: Homeless female lone parents in post-Olympics, austerity East London. *Cities* 76: 43–51.

Watt, P. (2018b). 'This pain of moving, moving, moving': Evictions, displacement and logics of expulsion in London. *L'Année sociologique* 68 (1): 67–100.

Watt, P. and Minton, A. (2016). London's housing crisis and its activisms: Introduction. *City* 20 (2): 204–221.

Weeks, K. (1998). *Constituting Feminist Subjects*. London: Verso.

Winton, A. (2007). Using 'participatory' methods with young people in contexts of violence: Reflections from Guatemala. *Bulletin of Latin American Research* 26 (4): 497–515.

Index

A Feminist Urban Theory for our Time: Rethinking Social Reproduction and the Urban,
First Edition. Edited by Linda Peake, Elsa Koleth, Gökbörü Sarp Tanyildiz, Rajyashree N.
Reddy & darren patrick/dp.
© 2021 John Wiley & Sons Ltd. Published 2021 by John Wiley & Sons Ltd.